Supply Chain Processes

Supply Chain Processes introduces readers to the view that genuine supply chain competitive advantage is achieved via supply chain excellence, which in turn is underpinned by supply chain *process* excellence. The ultimate supply chain goal is to deliver the right product, of the right quality, in the right quantity, to the right place, at the right time and for the right cost. The author identifies four core supply chain processes – *strategy, design, execution* and *people* – to which the process improvement techniques explained can be applied to ensure superior supply chain performance.

Key topics are addressed, including supply chain risk, resilience, sustainability, challenges confronting modern-day supply chains and developing an elegant supply chain design. Each chapter starts with a section that explores learning opportunities and is followed by detailed chapter content. Carefully explained examples are provided, as well as end-of-chapter case studies, review questions and written assignments. A full suite of teaching aids is also provided for lecturers and tutors.

Providing both management expertise and technical skills, which are essential to decision-makers in the supply chain, this textbook should be essential reading for undergraduate and postgraduate students, post-experience students studying for professional qualifications, and operating supply chain leaders and supply chain professionals.

Online resources include chapter-by-chapter PowerPoint slides, tutorial exercises, written assignments and a test bank of exam questions.

Peter W. Robertson is Honorary Research Fellow at the University of Wollongong (UOW), Australia.

Supply Chain Processes

Developing Competitive Advantage through Supply Chain Process Excellence

Peter W. Robertson

Routledge
Taylor & Francis Group
LONDON AND NEW YORK

First published 2021
by Routledge
2 Park Square, Milton Park, Abingdon, Oxon OX14 4RN

and by Routledge
605 Third Avenue, New York, NY 10158

Routledge is an imprint of the Taylor & Francis Group, an informa business

British Library Cataloguing-in-Publication Data
A catalogue record for this book is available from the British Library

Library of Congress Cataloging-in-Publication Data
Names: Robertson, Peter W., author.
Title: Supply chain processes : developing competitive advantage through supply chain process excellence / Peter W. Robertson.
Description: 1 Edition. | New York City : Routledge Books, 2021. | Includes bibliographical references and index.
Identifiers: LCCN 2020054742 (print) | LCCN 2020054743 (ebook) | ISBN 9780367540074 (hardback) | ISBN 9780367540081 (paperback) | ISBN 9781003084037 (ebook)
Subjects: LCSH: Business logistics. | Competition.
Classification: LCC HD38.5 .R6334 2021 (print) | LCC HD38.5 (ebook) | DDC 658/.7—dc23
LC record available at https://lccn.loc.gov/2020054742
LC ebook record available at https://lccn.loc.gov/2020054743

ISBN: 978-0-367-54007-4 (hbk)
ISBN: 978-0-367-54008-1 (pbk)
ISBN: 978-1-003-08403-7 (ebk)

Typeset in Bembo
by Apex CoVantage, LLC

Access the Support Material: www.routledge.com/9780367540081

This book is dedicated to my family:
To my patient wife Jenniffer.
To our children and their spouses Daniel, Sarah, Luke, Annalise, Hayley and Jason.
And to our wonderful grandchildren Penelope, Axel, Piper, Pace, Havana and Dexter.
With love.

Contents

Acknowledgements

I would like to acknowledge the following mentors and friends who gave me the inspiration and determination to undertake and complete this book:

Professor Tim Coltman, Dean (Tauranga) and Professor of Innovation and Technology Management, Waikato Management School, University of Waikato, New Zealand.

Associate Professor John Terrence Flanagan (deceased), University of Wollongong.

Distinguished Professor Emeritus Martin K. Starr, Rollins College, Winter Park, Florida, and Senior Professor Emeritus Columbia University, NYC.

About the Author

Peter Robertson is an Honorary Research Fellow and Fellow of the University of Wollongong (UOW), Australia.

Dr Robertson has 39 years of industrial experience working for the Broken Hill Proprietary Company Limited (BHP) and BlueScope Steel. Roles he fulfilled during this time included Unit Metallurgist, Rolling Mill Superintendent, Group Manager Planning and Manager Planning and Business systems. His last role was as Vice President Operations Planning within BlueScope Steel.

He has three years of lecturing experience at UOW where he lectured on Supply Chain Management (SCM), Logistics, Procurement and Quantitative Analysis.

Dr Robertson also spent five years consulting including assignments related to strategy and structure, business process improvement, value-stream mapping, project readiness and information systems design. In his consulting role he has worked with companies such as Woodside Petroleum, Port Kembla Coal Terminal, Port Kembla Port Corporation, Vale (Brazil), Anglo American, Chalco, BHP Iron Ore, Australian Nuclear Science and Technology Organisation (ANSTO) and Hatch Associates.

Dr Robertson obtained both his master's degree (Logistics and Operations Management) and PhD (SCM and Logistics) from UOW.

Whilst still engaged in research, Dr Robertson is more focused now on the writing of three SC and logistics related books i.e. SC Analytics (2021), SC Processes (this one) and SC Leadership (due 2022).

Foreword

It is a pleasure to compose the foreword for Dr. Peter W. Robertson's second textbook, in a series of three supply chain (SC) textbooks, titled: *Supply Chain Processes – Developing Competitive Advantage through Supply Chain Process Excellence.* It does indeed represent a continuation of the highly relevant and well-written supply chain management treatise conceived and executed by its author. If you don't know what a treatise really means, it is worth looking it up because that term truly applies in this case. The promise of three related texts is now accomplished two-thirds of the way without diminishing the power of the author's ideas and the clarity of his writing.

Dr. Robertson's stated intent is to write three texts in this compelling and ambitious supply chain management series: Supply Chain Analytics (completed), (2021), Supply Chain Processes (this one), (2021) and Supply Chain Leadership (due 2022).

True to his promise, Dr. Robertson presents in his second curtain call, a really detailed deep dive into supply chain processes. You know that you are going 'in depth' when you examine levels 4, 5 and 6 sub-processes in detail. Quoting from Chapter 3:

> In setting a competitive SC strategy, it is important to at first understand the reality of the situation confronting the SC. To do this, it is necessary to conduct an analysis of the business environmental conditions that exist. Some refer to such an analysis as an 'as-is' situation appraisal.
>
> Fortunately, several tools exist to help SC leaders and SC professionals with such an analysis. Three such tools are presented next: a political, economic, social, technological, legal and environmental (PESTLE) analysis; a strengths, weaknesses, opportunities and threats (SWOT) analysis and a vision elements analysis (VEA).

In fact, this remarkable scope is much more than a deep dive. It is a structural achievement, resulting in both foundational (solid) achievements and astronomic perspectives (the largest systematic viewpoint).

The connection between overall supply chain excellence and supply chain links and sub-link process excellence is woven successfully throughout the book. I have already pointed out that to understand supply chain complexity, the systems concept is fundamental. In this case, it is supported by the inclusion of a hypothesis-based supply chain management theory which unites the entire systems concept of supply chain management in a logical way. Dr. Robertson receives my applause and admiration for attempting and successfully delivering this long overdue theory.

The second volume contains not only detailed descriptions of supply chain processes that matter the most to supply chain excellence but it is also peppered with practical, easy-to-follow examples and relevant case studies that provide clear pathways for others to follow.

A unique feature of this book is its differentiation of the socio-technical dimensions of supply chain management. This includes the identification of typically confronted obstacles and barriers (especially the socio-barriers) to the obtainment of genuine supply chain process excellence and how to go about overcoming them.

Importantly, the book addresses, head-on, the issue of supply chain sustainability and devotes an entire chapter to this vitally important topic. The chapter includes how to specifically go about responding successfully to this growing supply chain challenge. At the same time, Chapter 7 can be readily seen as a stand-alone monograph that spells out antecedent aspects of sustainability that eventually translated to the triple bottom line concept. Peter W. Robertson reveals the sense of history that shows why sustainability requirements were not invented the day before yesterday. They have always been there, but the time to turn on the spotlight has now come.

Assurance is another high priority topic addressed in the book. This includes how to assure the relevance and quality of developed supply chain strategies and design. Understanding the assurance of on-time, on-budget, and to-promise delivery of supply chain projects is essential.

The author makes the point repeatedly throughout the book that supply chain management and supply chain processes specifically are a highly interconnected system, a gestalt. As an analogue, we have no difficulty comprehending that an automobile cannot function if missing key mechanical or electrical parts or having no fuel or lubricants. In the same way, a supply chain cannot function properly with missing or poorly performing processes.

Who needs volume two? I do, but it will also substantially add value to the respective portfolios of academics, students, supply chain leaders, practitioners and partners. It will also constitute a valuable reference source for engineers, information, communication and technology (ICT) professionals and business leaders. May I also suggest that many governmental politicians, no matter what their stripe, would do well to spend time with both volumes one and two of this emerging treatise.

The pedagogy of this book is excellent. Each chapter starts with a section that explores learning opportunities and provides detailed chapter content. It also provides carefully explained examples, as well as end of chapter case studies, review questions and written assignments. A full suite of teaching aids is also provided for lecturers and tutors. These features make the adaption of this text to an on-line learning medium really straightforward.

In conclusion, I commend this book to the discerning reader looking to find an in-depth description of supply chain processes and how they work. Further, readers wanting to understand the connections between supply chain process excellence, overall supply chain excellence, competitive advantage and stakeholder value-add should review this text. What we have in this book is a whole-of-system type roadmap which, if followed diligently, will result in substantial benefits to all supply chain partners and participants involved, as well as to all students and managers of supply chains. Might I point out that all living souls are in fact constantly interacting with their own personal and critically important supply chains.

I praise Dr. Peter W. Robertson for delivering his second masterful contribution to our understanding of how global supply chains work. We see that at last, after many years, the horizon is clearing, and we can see what is necessary to create and keep supply chains competitive in a demanding, dynamic and continuously changing world.

Martin K. Starr
October 25, 2020
Professor Emeritus, Columbia University and Rollins College

Preface

For many supply chain (SC) professionals, the ultimate SC goal is to deliver the right product, of the right quality, in the right quantity, to the right place, at the right time and for the right cost.

There are a lot of 'rights' in the previous statement, and it is the definition of those 'rights' that gives insight to the concept and practice of SC excellence.

SC excellence implies offering attractive products and services to customers whilst achieving superior service delivery to those same customers, including on-time and error-free deliveries with shorter order lead times, responsiveness to changing customer requirements, flexibility/agility to changing marketplace conditions, lower end-to-end SC inventories and lastly, decreasing total SC costs. These are all crucial competitive issues for every operating SC.

A core precept of this book is that SC excellence is underpinned by SC process excellence. The four SC processes that are used throughout the book and that have been determined by the author's research as key to SC excellence are:

- i SC strategy
- ii SC design
- iii SC execution
- iv SC people

Modern-day SCs face many uncertainties, considerable volatilities, growing complexities and definite ambiguities and fuzziness. This environment of instability and unpredictability makes it harder for SC professionals to do their jobs well and to lead and manage high-performing SCs.

In addition to dealing with this so-called VUCA (volatile, uncertain, complex and ambiguous) environment, SC leaders and professionals have to respond proactively to the important issue of sustainability. The three sustainability goals of social, environmental and economic performance simply have to be responded to appropriately and the performance on each goal progressively lifted. Taken in concert, this is no small challenge.

The evidence is compelling, however, that it can be done, with companies such as Cisco Systems, Colgate-Palmolive, Johnson & Johnson and Schneider Electric all leading the way with superior SC practices and levels of performance.

This book, therefore, is a study of the underlying SC processes that enable such performance excellence under such testing circumstances. The book provides not only a detailed description of the four key SC processes but it also provides practical examples of how to go about improving the performance of each of the four SC processes and the many SC sub-processes.

The obtainment of SC process excellence, thus, is the key to overall SC excellence and the development and maintenance of competitive advantage. Such outcomes are essential for SC growth and long-term survival.

Pedagogy

The pedagogy of the book has six main features:

i An opening section for each chapter explaining the learnings designed for the chapter
ii Detailed chapter subject matter description followed by key measures of success and practical examples illustrating how to go about achieving such success
iii Tables, diagrams and actual example vignettes to reinforce the concepts and methods covered
iv End of chapter case study
v End of chapter review questions and written assignment tasks
vi A complete set of supplementary materials, including lecture slides, exam questions and tutorial exercises

Book audience

The book has been written for three main audiences within the disciplines of operations management, supply chain management and logistics:

i Undergraduate and postgraduate students
ii Working people or mature-aged persons wanting to upgrade their qualifications such as the 'Supply Chain Leadership Professional Degree' offered by the UK's Institute for Apprenticeships and Technical Education (www.instituteforapprenticeship.org)
iii Operating supply chain leaders and supply chain professionals

1 Introduction and context

1.1 What you will learn in this chapter

The chapter firstly introduces the development of Supply Chain Management (SCM) including typical Supply Chain (SC) types and the complexities involved in the management of modern-day SCs.

Next, the four key SC processes used extensively throughout the book are listed.

Supply Chain Excellence (SCE) is then introduced along with the complications of managing for SCE in the modern-day business environment of increasing *Volatility, Uncertainty, Complexity and Ambiguity* (VUCA).

The context of the book is then described, including how the development and application of effective SCM methods and practices over recent decades have delivered market leadership to those companies embracing the methodology.

An introduction to the United Nations (UN) Sustainable Development Goals (SDG) is then provided, including a listing of the SDG most relevant to SCM.

The objective and value-add of the book is described, followed by a pertinent history of SCM presented as a case study.

The chapter ends with review questions and written assignment options.

1.2 Introduction to supply chain management and supply chain excellence

Supply Chain Management (SCM) has developed quite rapidly in a formal way since the early 1980s when the term was first used in literature (Laseter and Oliver, 2003).

In an informal way, however, SCM has been evolving for centuries. The ancient Egyptians, for example, utilised SCM in the building of their pyramids. By default, they were managing supply chains which involved quarrying large rectangular blocks of rock and transporting them to a pyramid construction site – such as Giza on the outskirts of Cairo – and then erecting them into the famous polyhedron shape. These were major supply chain activities, building royal tombs involving thousands of people mining and transporting large stones, followed by pyramid construction carried out over a one-thousand-year period, over 4,000 years ago (c. 2649 BCE to c. 1640 BCE).

Every enterprise since that time, involved in the sourcing of materials and services, transforming them in some way, transporting them to an end consumer and working with partners along the way, has been undertaking some form of SCM.

Over those many years of informal use and recent decades of formal development, SCM activities and management techniques have become quite complex and, in some

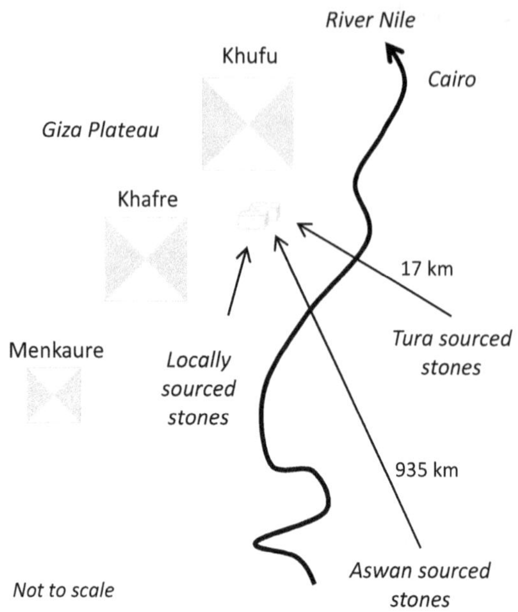

Figure 1.1 Giza Pyramids Main Material Supply Chain

cases, considerably advanced technically. Additionally, SCM techniques are varied and there is no 'one-size-fits-all' as an approach that may work well in one particular supply chain may not work so well in a different chain.

SC requirements also vary dependent upon the nature of the chain being operated. Examples include military SCs, humanitarian relief SCs (e.g. in response to disasters), critical health supplies (e.g. blood, body parts, medicines, vaccines, protective equipment), perishable goods supply chains (sometimes called 'cold-chains', mostly used for perishable foods but not exclusively), food SCs (perishable and non-perishable), fashion SCs (e.g. apparel, clothing), energy SCs (e.g. gas, oil, electricity) and mass transit SCs (e.g. planes, trains, trams, buses and ferries).

For experienced and competent SC professionals directly involved in such challenges and SC advancements, the proper design and running of SCs is obvious and the selection of the best methods by which to manage a particular chain and how to go about developing and sustaining necessary levels of SC excellence, whilst complex, are all eminently doable.

However, for outsiders, for some senior managers and especially for new starters, the whole SCM issue can be not only confusing but also confronting. Worse, if a SC operative is unprepared and/or unsupported, SCM concepts and associated challenges can be quite overwhelming, particularly when things go wrong. This oftentimes results in high levels of stress and sometimes in burnt careers.

So just what does it take to apply SCM well? How does a single enterprise or a network chain of enterprises go about achieving, in a sustainable way, distinct competitive advantages via the application of modern SCM principles and techniques?

The *purpose* of this book, therefore, is to answer such questions, to capture and record decades of SCM learnings and to explain them in practical and connected ways with examples of how they can be applied to lift SC performance to levels that achieve optimum outcome results for a given set of SC circumstances.

The foundation platform for this book is a four-process (four-construct) SC framework. The four key SC processes are:

i SC *strategy*
ii SC *design*
iii SC *execution* (reliable delivery of customers' value propositions)
iv SC *people*

This framework, derived empirically and tested via research, basically represents a modern SCM system; that is, it includes all of the critical elements necessary to run SCs well (competitiveness through excellence). Each process and its sub-processes are explained in detail throughout the book.

This book, thus, has been written primarily for *students* of SCM and SCM *professionals* who are looking for a reliable 'how-to' guide to developing and sustaining SC excellence with a strong emphasis on describing and explaining the 'why' for each 'how-to' so presented.

The book can also be utilised by SC leaders and academics wishing to present a process-based SCM subject.

The author has empirically derived the approaches to SCM recommended in this book over a combined period of some 40 years. The author has also proposed a *Supply Chain Theory* based on that experience and knowledge. The SC theory is presented in Chapter 2.

Supply Chain Excellence (SCE) is about building the capability to meet ever-changing customer needs, wants and desires. Such capabilities include the continuous improvement of customer value propositions *and* delivering them faultlessly *whilst* carefully managing all SC inventories and costs. SC capabilities, thus, are key enablers of long-term SC competitive success and survival.

Importantly, SCE is also about achieving long-term SC competitive success and survival in a responsibly sustainable way as discussed in detail in Chapter 7.

A significant *complicating factor* in the quest for SCE, of course, is a SC operating environment exhibiting increasing levels of *Volatility, Uncertainty, Complexity and Ambiguity* (VUCA). Low probability, high consequence events, for example, can have devasting negative impacts on SCs. Sometimes referred to as 'black swan' events, recent examples include the 2004 Indian Ocean earthquake and tsunami, the 2011 Japan (Fukushima) earthquake and tsunami and the 2020 COVID-19 pandemic. Techniques that can be used by SC professionals to better manage such VUCA and black swan realities are included in Chapter 2.

The *compelling reason* for the use of this book by both SC professionals and students is that it offers an approach to SCM that is more *complete* than previously available. That is, the competent application of methodical and tested SC practices to the research-derived four SC processes that matter the most to SC performance outcomes, by skilled and experienced people, is the key to achievement of sustainable SC excellence.

The four SC processes in question are shown at the top of Figure 1.2: SC *strategy*, SC *design*, SC *execution* and SC *people*. Coordinated improvement to this core set of SC processes leads to improved SC technical and cultural performance which in turn underpins sustainable SCE.

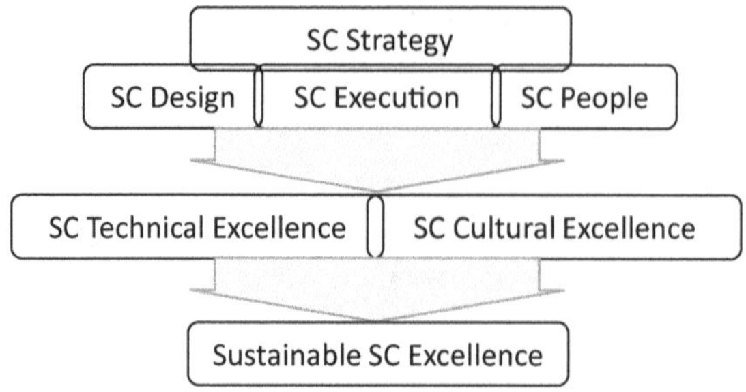

Figure 1.2 Four Crucial SC Processes as Enablers of Sustainable SC Excellence

SC technical excellence implies the capability to perform required SC functions safely and competently, thus ensuring timeliness, responsiveness, flexible order lot size, delivery reliability, suitable quality, inventory management and cost management. It also includes growth and renewal of the SC resource base.

SC cultural excellence implies many aspects including leadership quality, quality of relationships, commitment, determination and motivation of personnel, active personnel role networks, their interactions, collaboration and information sharing practices, levels of entrepreneurship, an external focus, an opportunity-seeking, courageous and energetic ethos, initiative, good business acumen, a fast pace of learning, competent management of risk and ecosystem awareness (What is changing? What is the likely impact? How should we respond? How might new opportunities be realised?).

A good example of an organisation that has developed and sustained SC excellence is the Spanish-based fashion firm Zara. Over a period of years, Zara has developed both a low-cost and quick-to-market model that has provided competitive advantage to Zara and its SC partners. Zara utilises worldwide materials sourcing, postponement techniques (e.g. garment dyeing instead of yarn dyeing) and advanced planning and scheduling of its production runs that is linked to point-of-sales (POS) data (allowing it to put on shelves the next day what customers are buying today). As a result of COVID-19, Zara, like many retailers, relies more heavily on a digital supply chain, including on-line sales.

SC professionals need to know how to develop and sustain such levels of excellence for their particular circumstances. For example, SC professionals might ask: 'What are the specific technical and cultural excellence attributes that are relevant to our SC?' 'What are the levels of excellence (targets) that we need to achieve on each of them and by when?' 'How do we go about it?' 'How do we develop and sustain such levels on such attributes?' These are the questions answered in this book.

The aim of this book is to help develop awareness and understanding of all of the components shown in Figure 1.2 such that students and SC professionals who are adequately trained in the book's principles can indeed help achieve sustainable SCE in practice and thus deliver real competitive advantage to their organisation(s).

SCE, however, can only be achieved if excellence is the performance standard of underlying SC processes. The author's research has identified four SC processes that are crucial to SC performance:

(i) SC *strategy*
(ii) SC *design*

(iii) SC *execution*
(iv) SC *people*

These four processes have been broken down into their level two, level three and level 'n' sub-processes. Key measures have also been identified for each. An example of this for the SC *people* process is shown in Table 1.1. An introduction to the four key processes is presented in Chapter 2, and full details of each process are presented in Chapters 3 to 6.

The content of this book, therefore, describes each of the four key SC processes in detail and, through the use of examples, explains how to manage the processes in order to obtain true process excellence and to identify how such a level of performance can be sustained.

Table 1.1 Level One to Three SC *People* Processes with Level Three Metrics

Level One	*Level Two*	*Level Three*	*Metrics Used for Level Three*
SC People	SC Leadership	Setting direction	An approved and up-to-date and comprehensive business plan exists
		Leveraging the potential of all resources	Effectiveness indexes exist for all personnel, physical assets and non-physical assets
		Leader selection	Leader technical suitability index Leader social suitability index
	Managerial System	Requisite-organisation	Approved organisational design that is aligned with organisational strategies exists Approved position descriptions for each position exist Approved remuneration policy exists Active effectiveness appraisal process exists Approved code of conduct exists Personnel inflow/outflow management system exists Training and assessment systems exist ICT support systems requirements definition exists
		SC people costs	Total personnel costs, cost/hour/classification Total costs per role type Costs per hour overall Costs per hour per role type Productivity index
	Organisational Culture	Principles, values and behaviours	Safety performance Number of appropriate behaviours Number of inappropriate behaviours Personnel turnover
	Knowledge Building	Open exploration and sharing of ideas	Active communities of practice exist, active recognition of best practice exists, technical support systems exist
		Collaborative learning	Number of active departmental & cross-functional improvement teams, number of active supply chain, cross-functional improvement teams, plus ratios of these to number of employees
		Advancing knowledge	Approved knowledge capture process exists, knowledge storage system exists, knowledge analysis and interrogation system exists, password protected, tiered access exists for all SC personnel

1.3 Context of supply chain management, supply chain excellence and competitive advantage

The context of this book is that the development and application of effective and advanced SCM methods and practices over recent decades have delivered market leadership for those companies leading the charge.

Companies such as Walmart, Dell, Unilever, DHL, Maersk, Amazon, Zara, Cisco Systems, 3M and BASF have confirmed by now that SCE is a genuine source of competitive advantage as it enables reliable fulfilment of customer order and service delivery expectations whilst effectively managing SC inventories and costs.

In short, SCE leads to higher levels of order delivery reliability, shorter order cycle times, lower supply chain inventories, lower total SC costs, greater SC responsiveness to customer and supplier changes, improved time-to-market performance for new product or service introductions and finally a proper product returns mechanism.

Such outcomes lead ultimately to competitive advantage, which manifests as higher economic returns for the organisations concerned. The top ten companies on Gartner's 'SC Top 25 for 2017', for example, showed an average three-year return on assets (ROA) of 14.4% (Ennis et al., 2017).

But are economic returns the only important measure of SC performance? The answer is no, especially not since the adoption of the UN 2015 SDG.

1.4 Supply chain sustainability

Prior to the release of John Elkington's book titled *Cannibals with Forks: The Triple Bottom Line of the 21st Century Business* (Elkington, 1997), most SC texts and articles concentrated on the many technical aspects of SC excellence required for success. With the release of the UN 2015 SDG, SC *sustainability* has become a mainstream SC topic.

Typical SC technical aspects, for example, included the specification and documentation of customer requirements, key SC measures, the imperative of customer service excellence, SC process design, process integration and process management, constraint management, operating methodology (e.g. 'Lean', JIT, Agile), order cycle times, batch sizes, product quality, planning and scheduling approaches, logistics, inventory management techniques, the application of ICT technology and SC costs.

The SC world, however, has changed since the adoption of the SDG and the whole issue of sustainability has very much widened the focus of SCM.

Whilst it could be argued that indeed *all* of the UN's SDG are relevant to SCM, it must certainly be the case that the following SDG are very relevant to SCM:

- i SDG #8 – Decent work and economic growth
- ii SDG #9 – Industry and infrastructure
- iii SDG #11 – Sustainable cities and communities
- iv SDG #12 – Responsible consumption and production
- v SDG #13 – Climate action
- vi SDG #17 – Partnerships for the goals

It is posited in this book, therefore, that a *technically centric* approach, whilst helpful, is *inadequate*. That is, comprehensive SC excellence is analogous to Yin and Yang, compromising

(i) *technical* excellence and (ii) *cultural* excellence. One element on its own creates imbalance; thus, both are necessary.

Today, therefore, SCE implies a level of overall SC performance that not only provides competitive advantage to the partner firms involved in operating a particular SC but that does so in a responsibly sustainable way.

1.5 The objective and value-add of this book

The objective of this book is to provide SC students and SC practitioners with a pathway which, should they follow it, will lead them and their organisation towards SCE and thus survival, growth and stakeholder value-add.

The true value-add of this book is the practical explanation via well-explained worked examples of how to go about achieving performance excellence on each of the SC processes most responsible for overall SCE.

The book provides not only a detailed description of the four key SC processes but it also provides practical examples of how to go about improving the performance of each process and the many sub-processes.

Further, included examples demonstrate how to apply various improvement methodologies such as Six Sigma, Lean Six Sigma and Analytics to each of the four key SC processes in order to obtain a level of process performance that represents genuine SCE.

Consider the case of an aluminium processing company that is looking to grow the output of its aluminium sheet. A complicating factor is that the company's material suppliers are distantly located along the SC and thus material transportation costs are high. The reason for the spread-out nature of this SC is that the processor located its facilities as close to the market as it could in order to reduce customer order cycle times. The aluminium smelters were located in regions where cheaper and reliable hydroelectricity was available, and the bauxite mines were located of course where the bauxite deposits lay.

This is a fairly classic SC design question as the processor must determine the best SC facilities location design that, within material and services supply constraints, satisfies customer service expectations whilst minimising all SC costs and the environmental footprint of the chain.

Understanding of the SC strategy and design processes and their sub-processes and of how analytic techniques can be applied to such processes in order to answer such facility design questions competently is one of the key undertakings of this book.

By careful and competent application of the knowledge explained in this book, therefore, SC professionals will be able to undertake original design and/or improvement initiatives that lead towards the ultimate goal of SCE for the SC they work in.

1.6 Chapter 1 case study – history of supply chain management and supply chain excellence attempts

SCM existed in reality long before the concept and practice were given that title. Indeed, military commanders for centuries have known that a large factor in the success or failure of their military campaigns was/is the strength and quality of troop and materials supply and transport. Armies for decades have referred to this as their 'logistics'.

Sun Tzu (544–496 BCE) a Chinese general who fought during the Warring States period following the decline of the Zhou dynasty set down his learnings which were captured, under his name, in the famous book *The Art of War* some 2,500 years ago (Minford, 2008). In this book, Sun Tzu explains:

> Without equipment, an army is lost;
> Without provisions, an army is lost;
> Without base stores, an army is lost.

Sun Tzu also explains the importance of communications, bringing equipment from home base but still foraging along the way and the dangers and cost of supplying armies at distance (Minford, 2008).

The Macedonians under Alexander the Great (356–323 BCE) relied heavily on speed and surprise for their military successes (Bose, 2003). Alexander's logisticians moved large numbers of people and materials across large distances and difficult terrains. They adopted such practices as, for example, supplying forward bases via ship and then dismantling the ship and carrying it overland until they reached the next navigable watercourse (Bose, 2003).

Bose (2003) proposes that Alexander mastered the art and science of military logistics, which can be summarised as:

1 *Keep it simple* – Only carry what is absolutely necessary and carry the same with the troops. Keep the logistics train as short and as light as possible
2 *Set up forward supply bases* – Tactical operations have often resulted in defeat because the SC of the advancing army becomes too long. Such stretched SCs experience increased supply lead times and run the risk of being cut by the opposing force(s)
3 *Plan far in advance* – Know where you are going in advance, study the terrain and know what the land along the planned route is capable of in terms of supplies
4 *Break into small units* – If forward supply bases are not possible, then break into small units and set each on a different route to the same destination. Along any individual route, send no more troops than the route can sustain
5 *Establish a single point of contact* – This is the single point of responsibility concept still practiced by many organisations today. Using this practice, everyone along the SC knows exactly who the 'go-to' person is

In similar fashion, the history of military logistics can be traced through the time of the Greeks, Romans, Arabs and various conflicts such as the Napoleonic Wars, American Civil War, World War I, World War II, up until the present time. Indeed, Lieutenant General William 'Gus' Pagonis was appointed as Director of Logistics (single point of contact) during the Kuwait/Iraq 'Desert Storm' operation of 1991 and applied most of Alexander's five principles of logistics to that military campaign.

The history of SCM, in so far as to how it came about for manufacturing and service organisations, had its genesis in the industrial revolution of the late 19th century. This period saw the development of interchangeable parts from approved suppliers and the beginning of mass production. Henry Ford was an early pioneer of not only mass production but also of vertical integration (e.g. Ford owned and operated iron ore mines and rubber plantations as well as his automotive manufacturing operations), which, by default, meant the recognition and management of extended SCs – albeit within a stand-alone company (Ashall and Parkinson, 2002).

World War II (WWII) saw a unique situation develop within the United States of America (USA) and the United Kingdom (UK) based manufacturing centres producing products for the war. In particular, aircraft factories suspended their 'push' production (or flowline) techniques that saw over activation of non-bottleneck resources, high work-in-process (WIP) and finished goods inventories, long lead times, poor quality, poor delivery performance and adversarial supplier relations and instead used a process that can best be described as lean SCs. That is, the necessity of the war demand and scarcity of materials saw lean SCs develop as the method of operation within the WWII factories which best satisfied the supply/demand requirements of the time. Of course, this manner of operation was not called 'lean' or 'supply chain management' back then, but the approach was the same as both modern concepts describe.

Following WWII, the intense wartime demand for military equipment and supplies significantly reduced. Underlying non-military demand was still strong of course, and manufacturing facilities in countries that still had productive capacity (i.e. not destroyed by warfare) were all very busy. However, instead of maintaining the lean approach that some facilities had used during the war years, they reverted back to (or continued to use) the push production (flowline) methods of the pre-war years.

The 1950s, however, was also a pivotal decade for lean, total quality management (TQM) and SCM. In that decade, W. Edwards Deming visited Japan on a number of occasions and so commenced the long, slow march of Japanese industry towards TQM. In addition, Kiichiro Toyoda (1894–1952) and Taiichi Ohno (1912–1990) of Toyota Motor Co. visited Detroit to study American car manufacturing techniques. Ohno, in particular, concluded that there must be a better way for Toyota to manufacture autos than he observed in Detroit, so he went home to Japan determined to develop that better way. Toyoda and Ohno did bring home one idea in particular that they wanted to use, and it was the shelf replenishment technique they observed in a US 'Piggly Wiggly' self-service grocery store. This replenishment technique would lead to Toyota's 'kanban' inventory management system, a key plank of just-in-time (JIT) and the Toyota Production System (TPS). Ohno developed the TPS (otherwise known as 'Lean Manufacturing') methodology over a period of some 35 years to include manufacturing, logistics, suppliers and customers (i.e. key supply chain partners). It is also instructive to note that when asked "What was your source of ideas for the TPS?" Ohno would reply, "*Today and Tomorrow* by Henry Ford" (1926).

Following WWII, Toyota faced quite different market and financial circumstances than did Ford and General Motors (GM) (Liker, 2004). For example, due to strong US domestic and international demand following the war, one US auto line could make up to 9,000 cars per month, whereas Toyota was making only one tenth that amount. Ford and GM were focused on economies of scale, high equipment utilisation rates and large batch sizes. Such mass-production methods, whilst generating high inventory levels and lengthy manufacturing lead times, did reduce cost/unit produced.

Toyota on the other hand was required to produce much lower volumes of different vehicles on the same production line. In addition, Toyota was short of cash and so needed short order lead times to reduce their cash-to-cash cycle time (time between paying vendors for materials and being paid by customers for finished products). Toyota, therefore, needed a manufacturing methodology that provided both flexibility and the ability to quickly deliver the right product, of the right quality, in the right quantity, in the right place, at the right time, for the right price.

Such requirements are the main drivers of SCM today, so here we have evidence of one set of SC performance invariants, the need to deliver desired and quality products with speed to the correct place, on-time, cost-effectively.

A point that will be revisited later in the 'SC *People*' chapter of this book is that Ohno undertook the development of the TPS against considerable resistance and criticism within Toyota. It is a mark of his tenacity that he achieved as much as he did given such early internal opposition.

The 1960s saw a continuation of both strong underlying demand for manufactured goods and predominance of 'Push' production methodology in the West. Ohno in Toyota continued to develop and refine the Toyota Production System (TPS).

The 1970s saw two oil shocks that not only inflicted severe economic impacts on many countries but also influenced business operating methodologies.

In 1973, the USA decided to re-supply the Israeli military during the Yom Kippur war. This led the members of the Organization of Arab Petroleum Exporting Countries (OAPEC – consisting of the Arab members of OPEC, plus Egypt, Syria and Tunisia) to proclaim an oil embargo. This first oil shock saw oil prices rise from US$3/barrel to US$12/barrel. This fourfold increase gave rise to obvious uncertainty and economic impacts, including a fall in demand for many manufactured products as consumer confidence wavered considerably.

The Japanese economy and corporate profits suffered substantially from this first oil shock. Significantly though, Toyota did not trade into the red as far as most Japanese companies and indeed recovered faster. This did not go unnoticed by the Japanese Government that soon requested Toyota to make the TPS approach known to other Japanese companies. The years of toil and debate and continuous improvement efforts by Ohno and his team were paying off.

The second oil shock occurred in 1979 when Iranian oil exports were disrupted with the fall of that country's Shar. In addition, the Carter administration in the USA decided to cease Iranian oil imports. This whole situation created panic, and the oil price rose from US$15/barrel in 1979 to nearly US$40/barrel by 1981 (Al-Chalabi, 1980).

The initial business response to such 'shocks' was to cut costs, cut back operating capacity and lay off labour. As time went on though, some savvy companies saw the oil shocks as a signal to more permanent change, so they began a search for better ways of operating. Led by leading-edge thinkers of the day such as Michael Porter (1980), companies began experimenting with the 'value-chain' concept, global competition and vertical integration.

In the 1980s, the successes Japanese companies were having with Total Quality Control (TQC) led to western organisations taking the whole quality issue much more seriously, and thus arose the Total Quality Management (TQM) approach which included, at last, a much-heightened focus and awareness of customers and their real needs.

Companies such as Motorola realised that they had to substantially reduce their order lead time and dramatically improve product quality if they were to remain competitive. So began the Six Sigma improvement process.

In 1984, the USA Apparel Industry Council was formed in an effort to rescue the local apparel industry from closure due to intense competitive rivalry from imports. This group initiated actions that lead to the development of the 'Quick Response' (QR) method of supplying customers. QR called for an end-to-end SC focus, a substantial reduction to overall apparel chain inventories, improvements to product (garment) quality and in particular, reductions to order cycle times.

In 1982, arising out of a project staffed by four Booz Allen and Hamilton (BAH) consultants working for the Swiss company Landis and Gyr, the term 'Supply Chain Management' was published for the first time in both the UK (Financial Times article) and

Germany (Wirtschaftswoche business magazine article). BAH partner Keith Oliver and BAH project manager Wolfgang Partsch were directly involved with this project.

In 1985, the Boston based consultancy Pittiglio, Rabin, Todd and McGrath (PRTM) used the term Supply Chain Management formally to describe a cross-functional set of integrated business processes (Cohen and Roussel, 2005).

In 1986, Goldratt and Fox introduced the concept of 'Synchronous Manufacturing'. This approach called for balanced product flows, buffered bottlenecks and control over the release of materials into the SC (the so-called drum, buffer, rope approach). Synchronous Manufacturing was later expanded (Goldratt, 1990) into the 'Theory of Constraints' – an approach for improving manufacturing and SC performance.

In 1989, Stevens described four possible stages of SC integration: (i) baseline, (ii) functional integration, (iii) internal integration and (iv) external integration. Organisational progression along this continuum in reality, however, proved to be very slow.

All of these 1980s actions set the stage for considerable advancement and greater understanding of SCM that was to occur during the 1990s.

The 1990s were the decade of legitimisation of SCM. Yes, there were still senior managers in some companies who would not even use the term (the whole concept of end-to-end SCM presented a considerable threat to the functional silo chiefs in classical hierarchical organisations – more on this later in the 'SC People' chapter); however, after such successes as those achieved, for example, by the US grocery industry's 'Efficient Consumer Response' (ECR) initiative, the business improvements possible via the active application of SCM principles could no longer be ignored.

The 1990s saw many SCM advancements including the definition of the key SC processes 'plan', 'source', 'make' and 'deliver' by a joint PRTM and Xerox project team. These processes were the foundation set for the Supply Chain Operations Reference (SCOR) model developed in 1996 between PRTM, AMR and a number of their client companies. The Supply Chain Council (SCC) was set up in 1997 and charged with the ongoing development and improvement of the SCOR model (Cohen and Roussel, 2005).

The 1990s also saw the development of computer-based applications for SC processes such as sales and operations planning, master production scheduling, MRP-III, unit scheduling, transport planning and scheduling, inventory and warehouse management systems.

Of relevance also was the publication of the book *The Machine That Changed the World* by Womack et al. (1990). Arriving some 17 years after the first oil shock, this work basically explained the principles of the TPS under the label 'Lean'. Womack and Jones followed up in 1996 with their very successful book *Lean Thinking*. Many western companies then began experimenting with this so-called 'Lean' approach. Unfortunately, however, and as Liker (2004) identified, very few of the early western attempts at 'Lean' were sustained.

The 2000s was a time of maturation for SCM. Many, many SC initiatives were undertaken such as:

- i Dell's direct-to-market assemble-to-order PCs
- ii Amazon's Internet marketplace
- iii Apple's time-to-market process, supply chain planning and capacity management improvements
- iv Walmart's supplier engagement program (e.g. Proctor and Gamble)

v Hewlett Packard's focus on engineering excellence and rapid product introductions
vi US apparel industry's use of 'Quick Response'
vii US grocery industry's use of 'Efficient Consumer Response'
viii Zara's very successful postponement strategy and production planning based on live point-of-sale data capture

Considerable research into SC method and strategies were conducted during the 2000s, and numerous publications of SC-related articles occurred. Notable books published include:

i *Design and Managing the Supply Chain* (Simchi-Levi et al., 2000, 2003)
ii *Adapt or Die* (Heinrich and Betts, 2003)
iii *Supply Chain Excellence* (Bolstorff and Rosenbaum, 2003)
iv *Logistics and Supply Chain Management* (Christopher, 2005)
v *Strategic Supply Chain Management* (Cohen and Roussel, 2005)
vi *Principles of Supply Chain Management* (Wisner et al., 2005)
vii *Supply Chain Management* (Lambert, 2006)
viii *Living Supply Chains* (Gattorna, 2006)
ix *Supply Chain Logistics Management* (Bowersox et al., 2007)
x *Supply Chain Management* (Fawcett et al., 2007)
xi *Supply Chain Management* (Chopra and Meindl, 2007)

The decade from 2010 to 2020 was a period of substantial economic, political, social and environmental change. Such change drove a shift in SCM priorities. The top SCM priorities for the decade became:

i SC security and resilience. This priority was highlighted especially following the 2011 Japan earthquake and tsunami and the 2020 COVID-19 pandemic. Such shocks substantially disrupted SC operations, leading to supply shortages, delivery disruptions, volatile demand levels and subsequent SC economic impacts
ii A heightened focus on meeting changing customer requirements and the need for high levels of customer service performance
iii The need to bring new products and services to market quickly and reliably
iv Compliance with government and societal requirements especially around the drive for sustainability, including environmental footprint minimisation
v Technological advancements (especially the emergence of Cloud Computing, increased SC visibility, Blockchain, Supply Chain Analytics [SCA], Big Data Analytics [BDA], data mining and fully digital supply chains)
vi Volatile raw material prices and the need for high supplier performance
vii Importance of SC partner collaboration enabled by Cloud Computing
viii Ongoing need for SC cost competitiveness and productivity improvements

Notable SC publications for the decade from 2010 to 2020 include:

i *Logistics and Supply Chain Management* (5th Edition, Christopher, 2016)
ii *Operations Management for Business Excellence* (Gardiner and Reefke, 2019)
iii *Supply Chains in Action* (Wilding, 2019)

A key objective of this book is to distil such historical knowledge and to provide guidance (a pathway) to supply chain students and practitioners on how they can go about building SCE for the SCs they operate in.

1.7 Review questions

- i List six different SC types. How might the requirements of each differ?
- ii The foundation platform for this book is a four-process (four-construct) SC framework. What are the four key SC processes and why do you think they matter?
- iii SCE is mentioned many times in this chapter. Just what is SCE and why is it important?
- iv What six UN Sustainable Development Goals are very relevant to SCM and why so?
- v Sun Tzu is quoted as stating: "Without equipment, an army is lost; without provisions, an army is lost; without base stores, an army is lost". What do you think this means? Is it still relevant today? If yes, give some examples.
- vi What were the five SC traits that Alexander the Great mastered? Are any (or all) of them still relevant for SCM today?
- vii Did Toyota's Ohno develop all the practices of the Toyota Production System (TPS or Lean) from first principles or did he 'borrow' ideas from elsewhere? If he borrowed ideas from elsewhere, what are some examples?

1.8 Assignment topics

(Each to be 3,000 words minimum or about three A4 pages single spaced. Tables and figures are not to be included in the word count.)

- i 'Companies such as Walmart, Dell, Unilever, DHL, Maersk, Amazon, Zara, Cisco Systems, 3M and BASF have confirmed by now that Supply Chain Excellence (SCE) is a genuine source of competitive advantage as it enables reliable fulfilment of customer order and service delivery expectations whilst managing Supply Chain (SC) costs'. Discuss this statement. From your own research of these companies, do you agree that they are SC leaders? Why so? Are there other companies that you would consider as SC leaders? If yes, then what is it about their manner of operation and/or results they achieve that leads you to believe so?
- ii 'SCE leads to higher levels of order delivery reliability, shorter order cycle times, lower in chain inventories, lower total SC costs, greater SC responsiveness to customer and supplier changes, improved time-to-market performance for new product or service introductions and finally a proper product returns mechanism'. From your own research, identify at least three companies that are achieving such results. Explain the main SC features that each of your chosen companies utilise.
- iii From the results of your own research, provide a description of the six UN SDG that are most relevant to SCM. How might SCM need to change to achieve high levels of alignment with these six SDG? Are there other of the UN's 17 SDG that SCM should also embrace? If yes, which one(s) and why so?
- iv The SC 'world' has changed substantially over the last decade. What were those changes, how did they impact SC customers and SC operations and what should SC professionals do to best respond to them?

1.9 Chapter 1 references

Al-Chalabi, F. (1980). A Second Oil Crisis? *OPEC Review*, 4(4), pp. 8–19.

Ashall, D., Parkinson, B. (2002). Leaning Towards Agile. *Manufacturing Engineer*, 81(1), pp. 27–32.

Bose, P. (2003). *Alexander the Great's Art of Strategy*. Sydney: Allen & Unwin.

Cohen, S., Roussel, J. (2005). *Strategic Supply Chain Management*. New York: McGraw-Hill.

Elkington, J. (1997). *Cannibals with Forks: Triple Bottom Line of 21st Century Business*. Oxford: Capstone Publishing Ltd.

Ennis, K., Romano, J., Aronow, S. (2017). The Gartner SC Top 25 for 2017. Available at: www.gartner.com/en/documents/3728317#1392532660 (Accessed September 2017).

Ford, H. (1926). *Today and Tomorrow*. Garden City, NY: Doubleday, Page and Co.

Goldratt, E. (1990). *Theory of Constraints*. Croton-on-Hudson, NY: North River Press.

Laseter, T., Oliver, K. (2003). When Will Supply Chain Management Grow Up? *Strategy+Business*, 32.

Liker, J. K. (2004). *The Toyota Way*. New York: McGraw-Hill.

Minford, J. (2008). *Sun-Tzu The Art of War*. Suffolk: Penguin Classics.

Porter, M. E. (1980). *Competitive Strategy*. New York: Free Press.

Stevens, G. C. (1989). Integrating the Supply Chain. *International Journal of Physical Distribution and Materials Management*, 19(8), pp. 3–8.

Womack, J. P., Jones, D. T. (1996). *Lean Thinking*. New York: Simon & Schuster.

Womack, J. P., Jones, D. T., Roos, D. (1990). *The Machine That Changed the World*. New York: Free Press.

2 Seven key challenges confronting modern-day supply chains

2.1 What you will learn in this chapter

The chapter starts with an explanation of the typical complications existing in modern-day SC operating environments. In particular, this involves increasing levels of *Volatility, Uncertainty, Complexity and Ambiguity* (VUCA).

Next, the *seven specific SC challenges* that such an environment presents are explored.

Every challenge requires a response, so four *response essentials* designed to deal with the seven identified challenges are described.

Next, *Supply Chain Excellence* (SCE) and the underpinning SC processes that enable its delivery are described in more detail.

A *Theory of Supply Chain Management* (SCM) is then proposed and identified as the logic of this book.

Lastly, a case study involving the quest for *Supply Chain Resilience* (SCRes) within a large Indian automotive and farm machinery manufacturer is presented. Importantly, at the end of the case study, a '*SC Resilience Builder*' is proposed for use by SC professionals.

2.2 Contemporary supply chain operating environment

As mentioned in Chapter 1, modern-day SCs face increasing levels of *Volatility* (e.g. fluctuations to demand levels, supply availability and/or prices), *Uncertainty* (e.g. COVID-19 may or may not diminish naturally, a vaccine may or may not be found), *Complexity* (e.g. a given situation is multi-faceted, has many variables and interconnections) *and Ambiguity* (e.g. relationships are unknown for sure, conflicting evidence exists). The acronym VUCA has been coined to identify this type of environment.

Such a business setting creates distinct challenges for all SC operatives, and it would be tempting for them to throw their hands up and say, "It's all too hard!" Such a response, however, cannot be allowed to happen as it would lead to ruin. How then should SC operatives best respond to such challenges? This chapter begins to answer that question, and the remaining chapters expand on how to go about it in detail.

Firstly, *seven specific SC challenges* that a VUCA type SC operating environment presents are outlined.

2.2.1 Volatile business environment

Volatility of the business environment SCs work within can be driven by a number of factors, including changes to:

i The customer base (number of, identity of, size, power and location)
ii Customer needs, wants, desires, buying preferences and order amounts

iii On-line buying patterns, especially when utilising an omnichannel marketing strategy that aims to provide the customer with a seamless and unified cross-channel experience
iv Products and services offered
v Processes and technologies used
vi Direct partners (suppliers, service providers, their category [strategic, bottleneck, commodity, leverage], the power they wield, their structure, relationship strength, specifications and performance on key measures)
vii Indirect partners that play an indirect but still supportive role in the movement, storage and transformation of products/services across the supply network
viii Competitors and competitive activity
ix Economic factors
x Chronic SC problems (e.g. constraint management, batch sizes, changeover times, product/service quality, product/service complexity, equipment reliability, process throughput variability, SC layout and geography, planning and scheduling effectiveness, availability of reliable information, red tape and bureaucracy)
xi Acute SC problems (e.g. earthquakes, tsunamis, floods, conflicts, insolvencies, strikes, material shortages, anti-dumping actions and pandemics)
xii People and personalities

2.2.2 Difficult to predict levels of competitive intensity

Competitive intensity is not easy to predict and varies with changes to:

i The degree of *differentiation* between SC competitors. For example, if competitors copy another SC's differentiation features, then rivalry between them will increase as customers realise little difference between their individual value offers
ii Reductions in competitors underlying *cost base* can and often does lead to increased rivalry as such cost reductions can be passed to customers as price reductions
iii *Market growth or decline*. Growing markets usually lead to reduced rivalry whilst market decline intensifies rivalry as competitors fight over shrinking demand
iv *Industry concentration*. High levels of industry concentration signify many competitors and has been described as a 'red ocean' situation (Kim and Mauborgne, 2005). Low levels of concentration signify few competitors or a so-called 'blue ocean'
v *Switching costs*. If a customer's switching costs (i.e. the cost of switching from supplier A to supplier B) reduce, then competitive intensity increases and vice versa

2.2.3 Degree of robustness of SC processes

SC processes that are fragile will fail when under stress, whereas SC processes that are robust will hold up under stress. SC process failure leads to performance failures such as stock-outs, missed deliveries, long lead times, poor quality, excessive inventories and/or cost excursions.

For example, consider the case of AlySmelt, an aluminium smelter. Because AlySmelt spent little time on its SC design process, it suffered stock-outs due to a supply disruption from its bauxite supplier.

In response, AlySmelt utilised SC Analytics (SCA) in order to predict likely future supply failures in terms of probability and scale. Using this information, AlySmelt decided on two key SC design features:

i A calculated level of bauxite *safety stock* to be held by its supplier at the mine
ii The implementation of a two-supplier strategy. This led to the development of a supply relationship with another supplier. Whilst the overhead of having to deal with two suppliers was increased, the reduction in likelihood of feed stock-outs more than compensated for the increased overhead cost. AlySmelt, thus, designed SC robustness into its SC

There are other events, however, that are much harder to predict. Typically they don't occur very often, but when they do occur, their impact is very substantial. Such events are referred to as low probability, high consequence events and are sometimes called SC shocks or 'black swan' events. Examples include violent storms, earthquakes, tsunamis, volcanic eruptions and pandemics.

Building SC resilience to such major shocks is not easy; however, it can be done. Ways of doing it are explored in Section 2.3.

2.2.4 Risks from use of SC improvement tools and practices

Over the past 70 years and certainly over the past 30 years, many SC improvement methodologies and tools have been developed and tested in practice. However, the results achieved from the application of such methodologies have been variable. Indeed, most attempts have ended in failure due to a combination of insufficient planning, unpreparedness, lack of commitment, resistance to change, change of leadership, lack of funding, changing priorities or the next 'flavour-of-the-month' initiative arrives. Examples of proven improvement methodologies include:

i The Toyota Production System (or 'Lean') as developed by Ohno (1988) and his colleagues between 1949 and 1975
ii Collaborative Planning, Forecasting and Replenishment (CPFR) process
iii Quick Response Manufacturing (QRM)
iv Total Quality Management (TQM), including the ISO 9001 Quality Management System
v Six Sigma
vi Lean Six Sigma
vii Supply Chain Analytics
viii Data Mining
ix Plan, Do, Check, Act (PDCA) continuous improvement cycle
x Theory of Constraints (TOC)

2.2.5 Supply disruptions

SC supplies can be disrupted for a number of reasons including:

i Suppliers' equipment failure
ii Suppliers' raw material shortages

iii Supplier insolvency
iv Competitor action
v Regulatory changes
vi Transport accidents and failures
vii Natural disasters
viii Human-made disasters
ix Strikes
x Anti-dumping actions
xi Customer demand overload

2.2.6 Difficulty in developing and retaining high quality and motivated staff

High quality and motivated staff can be and usually are the reason for the difference between SC excellence and SC mediocracy. There are many variables influencing staff performance and availability. Important examples include:

i Quality of SC leadership
ii The existence and use of reliable selection, induction training and assessment programs
iii A reliable and regularly used effectiveness appraisal process
iv The existence and use of a competent personnel management system, including a written position description for each role that clarifies specific job goals, expectations, relationships, accountabilities and authorities
v Clear signals to perform
vi Active promotion and usage of teamwork
vii A supportive, respectful and inclusive work environment
viii Level and appropriateness of the rewards and recognition system used
ix Staff poaching by competitors
x Age of the workforce
xi Economic conditions
xii The contribution of the work to the human condition
xiii Nature and variety of the work involved
xiv Behaviour and attitude of colleagues

2.2.7 The whole question of sustainability

Following the adoption of the UN 17 Sustainable Development Goals (SDG) in September 2015, the question of sustainability has been brought into very sharp focus. SCs are not exempt from the SDG expectations as at least six of the goals are very relevant to SCM:

i SDG #8 – Decent work and economic growth
ii SDG #9 – Industry and infrastructure
iii SDG #11 – Sustainable cities and communities
iv SDG #12 – Responsible consumption and production
v SDG #13 – Climate action
vi SDG #17 – Partnerships for the goals

Before the goals were endorsed by the UN, there is no doubt that many SCs were pursuing some, if not all, of the previously mentioned six goals. The expectation now, however, is that the six goals will rise towards the top (if not *the* top) of the SC agenda.

2.3 Responding competently to such an operating environment

An important starting point when considering how to respond to a VUCA type SC environment is to recognise that the previously mentioned seven SC challenges are interrelated.

For example, a downturn in an economy may well trigger an increase in competitor intensity (fight for share in a declining market). Fragile SC processes will be exposed in such a situation, typically resulting in SC performance failures. A SC halfway through the application of a SC improvement project may find funding for the project withdrawn due to a shortage of cash as a result of the downturn. Cancelling a project halfway through its execution most often results in wasted effort and funds for little or no performance improvement result. Perhaps even worse, quality staff involved with the project may feel so disappointed (even betrayed) that they exit the SC and sign-on with a competitor.

The interconnectedness between the seven SC challenges and SC performance is thus quite extensive, and the numerous linkages involved are illustrated in Figure 2.1.

In order to respond competently to the seven SC challenges, therefore, requires mastery of the four response essentials shown in Figure 2.2.

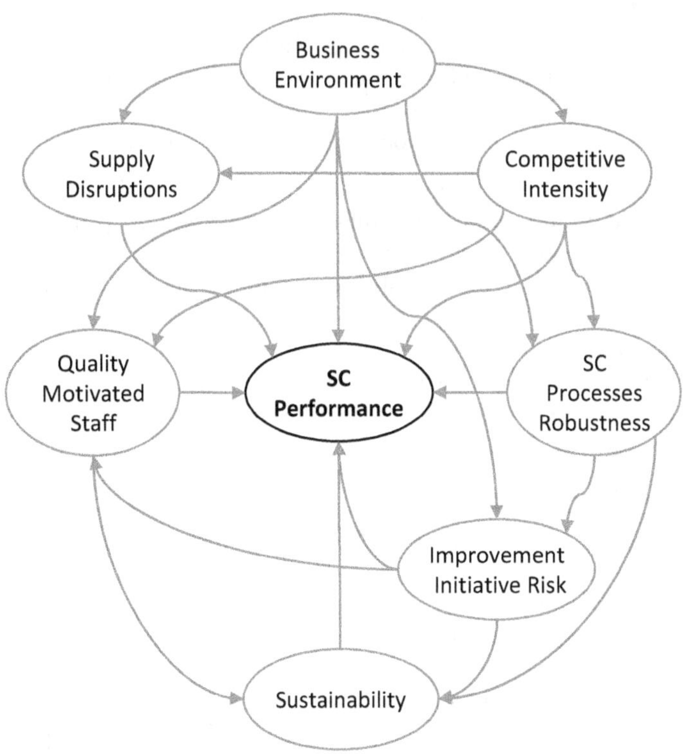

Figure 2.1 SC Challenges and SC Performance Interrelationships

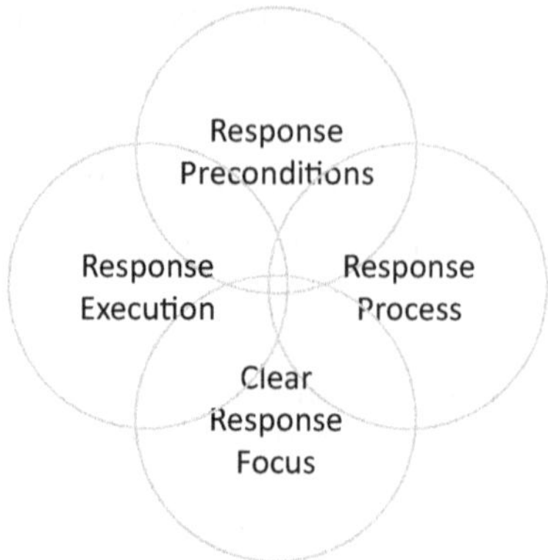

Figure 2.2 VUCA Type SC Environment Response Essentials

Taking the four response essentials one at a time:

i Response *preconditions* include:

  a Qualified, accredited, experienced SC leaders and SC professionals formed into a response team
  b Awareness of all of the interrelationships as shown in Figure 2.1
  c Detailed knowledge of the underlying SC processes involved
  d An appropriate toolkit, including proven SC improvement methodologies such as 'Project Readiness and Assurance', 'SC Shock Resilience Builder' and reliable tools such as SC Analytics (SCA)
  e Determination, persistence, focus, commitment and teamwork

ii Response *process* – This is essentially the roadmap or set of guidelines that the response team is to follow and includes a carefully developed response plan. The key elements of such a response process are shown in Figure 2.3

iii Clear response *focus* – Each SC challenge response team needs to very much focus on the specific challenge, the risk it presents, how to prepare for it, how to respond to it and the execution of that response. Such a level of focus, as illustrated in Figure 2.4, is crucial to response team success

iv Response *execution* – Unfortunately, many teams rush to execution before they have properly conducted the necessary levels of pre-execution planning. Invariably, this leads to execution failures. Execution should only be attempted, therefore, *after* the necessary planning activities are finalised and approved. Careful management of the lead times required to do this is therefore necessary

Successful execution also requires sustained and competent effort usually by a number of individuals working in teams. This means that team members may need to be encouraged to change their work behaviours, sharpen their competencies and lift their motivation levels.

Such an insight suggests that any execution process must include not only the technical activities that must be undertaken but also the cultural aspects that have to be observed in order to achieve success.

A recommended execution process is shown in Figure 2.5.

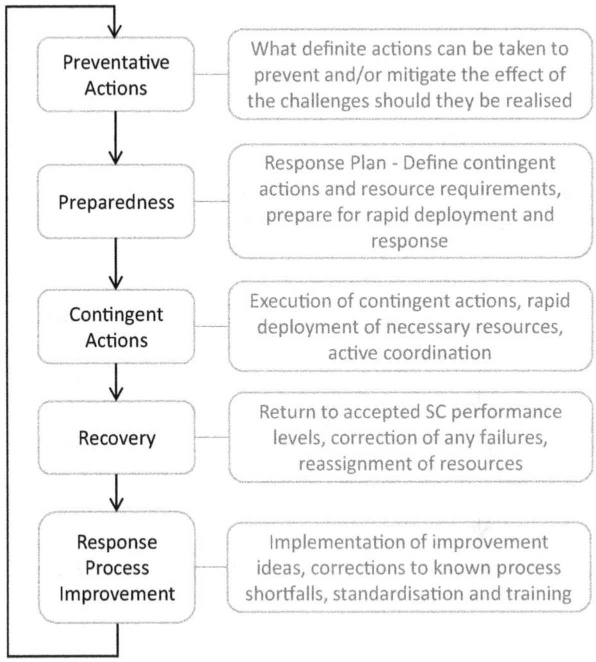

Figure 2.3 SC Challenges Response Process

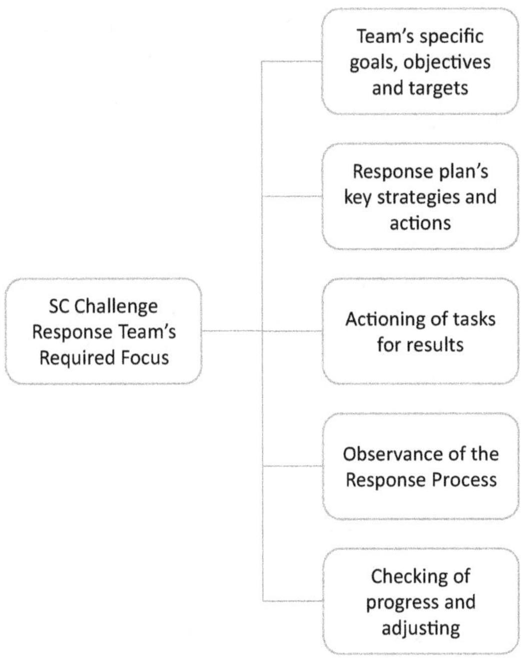

Figure 2.4 SC Challenge Response Team's Required Focus

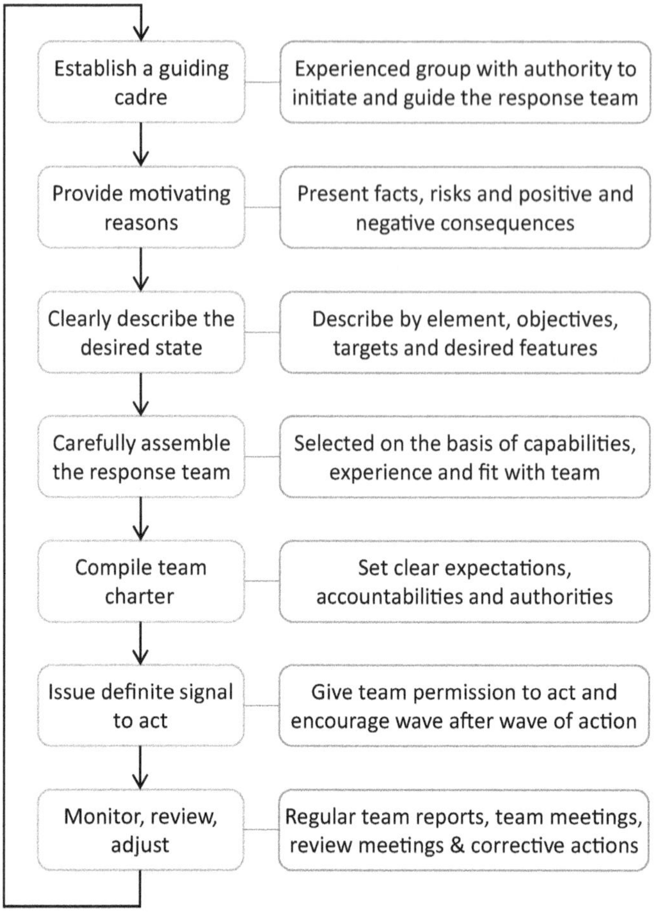

Figure 2.5 SC Challenge Response Execution Process

In concluding this section and, importantly, the previously mentioned SC challenge response process, whilst key to effective management of the environment confronted by modern-day SCs, *is not enough by itself*. This is because an even deeper level of capabilities is required to attempt not only a competent response to the seven SCM challenges outlined but also to the obtainment of SCE more broadly. Such capabilities provide the basis of the SCE aspects alluded to in Chapter 1. The next section, thus, explains the relationship between SCE and the four-key SC processes.

2.4 Supply chain excellence and the supply chain processes that underpin it

The obtainment of genuine SCE is a major undertaking for any SC. Before embarking on such a venture, therefore, it is important to understand the elements of SCE.

As alluded to in Chapter 1, SCE has two main elements: *technical excellence* and *cultural excellence*. Each of these elements is now described further.

Technical excellence – This is the capability to effectively manage the SC's technical requirements. Such technical requirements consist of:

i Detailed understanding and definition of customer needs, wants and desires. For example, smartphone suppliers such as Apple and Samsung need a clear understanding of just what it is that customers are looking for when considering which smartphone to purchase. Attributes such as dimensions, weight, colour range, damage resistance (scratching, bending, screen cracking), shock resistance (if dropped for example), water resistance, interface type and layout, screen resolution, battery life, device life, reliability, features (e.g. cameras, voice recorders, personal assistants), availability of attractive apps, interconnectivity, ease of use, availability of support, total costs, availability (supply lead time) and service-provider plans all need to be researched and understood from the customers' perspective

ii Product or service specifications and tolerances. Keeping with the smartphone example and considering just the lightweight case that houses the electronic components and display (and perhaps a keyboard), this case could be made from a range of materials including carbon fibre, resin (e.g. fibreglass), polymer (e.g. plastic) or metallic alloy. The smartphone supplier will need to decide which material to use based on customer preferences around weight, strength, durability, shock resistance and cost. Let us assume that a given supplier chooses a metallic alloy as most suitable. Again, there are numerous options that must be decided on: for example, what type of alloy (e.g. titanium/carbon/aluminium alloy), what chemical composition (e.g. aim % Ti, C and Al and the tolerances around each aim that are to be applied). Also, what maximum percentage of 'undesirable' elements is permissible (e.g. phosphorous, silicon, sulphur, nitrogen or hydrogen) and what desired mechanical properties are to be met (e.g. tensile strength, hardness, ductility, toughness). What is to be the manufacturing method used (e.g. powder metallurgy, casting, pressing or 3D printing), what type of surface finish is required and finally what are the key dimensions and their individual tolerances (deviation from aim that is acceptable – e.g. length, width, depth, thickness of casing material, curvature of edges and corners and overall weight). As can be appreciated from the previously mentioned description, this product and service specification task is formidable as each individual component for every product and service must be similarly researched, understood and defined

iii Definition of SC measures to be used, including targets per time period. Key *customer centric* SC measures are: (i) delivery (of customer orders) in-full, on-time and error free (DIFOTEF), the error free component denotes exemplary quality; (ii) order cycle lead time (time taken from customer placement of order to actual delivery of the order to the customer); (iii) customer responsiveness (time taken to respond to customer enquiries, to provide updates and support to customers, to respond to customer complaints and to respond to customers' changing circumstances) and (iv) customer flexibility (ability to flex production capacity up or down in accord with changing customer demand patterns). Setting up a relevant, end-to-end chain consistent and competent SC measurement and reporting system (including selection of measures, setting of targets and timing, collection of data, storage of data, analysis of data, reporting of results, visibility of information and any necessary follow-up corrective actions) is a crucial activity for SC professionals

iv SC *execution* process design (including value-stream mapping and agreed levels of SC process integration). A key activity of overall SC *design* is the design of the SC processes to be used. Process design can be carried out using flow-charting, stream flow-charting (where each stream represents an organisational function (e.g. research, marketing, production or SC)) or using algorithmic methods

Key SC *execution* processes include:

a Customer service and customer query management
b Order capture and order management
c Order cycle time management
d SC constraint management
e SC planning and scheduling
f Materials management (e.g. materials requirements planning [MRP] processes such as MRP III)
g Supplier management
h Order fulfilment and on-time delivery management
i SC inventory management
j Flow and logistics management

v SC information and communication technology (ICT) support systems such as order entry and order management, advanced planning and scheduling (includes MRP and constraint management rules), material tracking, supplier and customer relationship management systems (SRM and CRM), transport and warehousing management systems, SC information sharing and visibility systems

vi SC operating methodology (e.g. push, pull, lean, Agile). A key decision for SC leaders is to determine which SC operating methodology is most appropriate for their chain. Toyota, for example, runs a very 'Lean' methodology as imbedded in its Toyota Production System (TPS). Zara on the other hand runs a very agile SC as they have to be able to respond quickly to changing and fickle customer demand. As such, Toyota has a very strong focus on standardisation, low days of inventory, small batch sizes, quick change over times, kanban managed product flows, reduction of all forms of errors and waste and tight cost management. Zara rather, whilst exhibiting many of Toyota's SC features, has added emphasis on postponement, dynamic scheduling of production runs based on daily point-of-sales data and short product shelf lives (e.g. Spring, Summer, Autumn and Winter fashion cycles) driving the need for short time-to-market cycles for new products

vii SC flexibility and responsiveness management. SC flexibility denotes the ability to manage SC capacity to match changing customer demand and suppliers' supply patterns. Perhaps the best example of this is the managing of the critical transition from an old product model to a new one. Sales of the old model will fall off leading up to the launch of the new one. Companies, therefore, must work their capacity plans in close concert with price discounting on the old model. And market testing of the new product must be carefully carried out in order to better forecast likely sales of the new model. If this transition is poorly managed, the implications for the firms and certain employees involved can be substantial. Responsiveness management has everything to do with how quickly and efficiently responses to all customer and supplier issues are achieved. This includes responses to customer enquiries, customer order-entry cycle time, order delivery lead times, timeliness of customer updates, ease of and response cycle for all customer calls for support, response time to customer complaints and timeliness of invoicing. It also includes response to changing supplier and service provider circumstances

viii SC coordination process and mechanisms. As SCs become more global, their complexity increases on five main dimensions:

   a The number of 'actors' involved
   b The geographic distance travelled
   c The number of countries crossed, each with different bureaucratic rules, regulations, fees and time zones
   d The diversity of transportation modes (e.g. motorbikes, cars, trucks, front-end loaders, forklifts, conveyors, pipelines, barges, rail, ships and aircraft)
   e The diversity of cultures to deal with

   Such heightened complexity makes the task of SC coordination more arduous. Processes such as service level agreements, information sharing, information visibility, written role networks including clear outcome responsibilities, shared measures and rewards, shared personnel, shared planning, scheduling, product tracking and inventory management systems must all be put in place along the chain

ix SC cost management. Key elements of SC cost to manage include all material costs (suppliers, storage, transport, holding costs, systems and personnel costs) and order fulfilment costs (capturing orders, order entry, order management, order processing, transport, warehousing, delivery, invoicing, returns, systems and personnel costs). Using relevant benchmarks, competitive cost targets aligned with a SC's chosen basis-of-competition (e.g. innovator [Hewlett-Packard], first-to-market [Apple, Samsung], service/solution provider [IBM], technology leader [Intel]) need to be set for each element of the two main SC cost groups. Then actions to meet the targets need to be developed, assigned single-point responsibility, by when, and then implemented and tracked

Cultural excellence – Capabilities necessary to be able to evolve the SC in order to competently meet changing circumstances promptly. Such capabilities include:

i Leadership quality. Great leaders inspire great results (Collins, 2001). Examples of leaders who ran companies that substantially outperformed their competitors are Darwin Smith (CEO of Kimberly-Clark from 1971 to 1991), Colman Mockler (CEO of Gillette from 1975 to 1991) and George Cain (President of Abbott Laboratories from 1958 to 1962 and Chairman from 1962 to 1972). Leadership attributes exhibited by such CEOs that are essential for the success of any organisation or SC are:

   a Trustworthy – Leaders exhibit modesty and are self-effacing, mature, diligent, honest, respectful, open, inclusive and engaging. This is quite a hurdle in our modern age, and it will come as no surprise, thus, that genuinely trustworthy leaders are hard to find
   b Inspiring – The ability to not only paint and sell an inspiring vision of the future but also to set inspiring standards and to insist on the levels of discipline necessary to observe those standards
   c Passionate – An enduring determination and driven by a personally felt need to produce valuable results
   d Competent – Professionally trained, skilled, intelligent, knowledgeable and experienced
   e Present – Is physically and emotionally present, not unavailable (always travelling) or aloof. Is able to relate, shows genuine interest and contributes

f Constancy of purpose – This denotes an unwavering resolve to do what the leader is convinced (from knowledge and experience) must be done

ii Quality of relationships. Relationship quality stems from trustworthiness, listening, showing respect, a caring attitude, paying attention, regular contact, equitable sharing of credit and acceptance of responsibilities

iii The attributes of SC Personnel, including:

a Commitment – Actively engaged in the delivery of objectives and obligations
b Determination – Demonstration of resolve and courage in delivering required results and resilience and tenacity in the face of difficulties
c Motivation – Demonstration of drive and enthusiasm in undertaking actions to deliver goals without continuous supervision
d Energy – The necessary vigour to actually perform real work
e Responsibility – Acceptance of duties and the need to carry them out, acceptance of the need to deliver results and acceptance of onus when things do not work out as planned
f Initiative – Demonstration of resourcefulness and creativity when dealing with challenges
g Business acumen – In-depth knowledge of the business, its external environment, customers, suppliers, partners, competitors, people, processes, products and services
h Pace of learning – The ability to take in and take on new knowledge, new approaches, new products and processes, plus new opportunities and threats

iv Role networks of personnel, their interactions and information sharing practices. Role networks must be mapped out, regularly tested for performance and revised as necessary

v Creation of the people resource base. Stemming from the SC's basis-of-competition definition, a description of the required people resource base needs to be undertaken. Steps then need to be defined in order to put the desired people resource base in place

vi Growth of the people resource base. This includes a definition of which parts of the resource base need to be grown, on which dimensions, by how much and by when

vii Evolution/renewal of the people resource base. Essentially a definition of which parts of the resource base need to be evolved/renewed, by when and how

viii Recognition of opportunities and an energetic and active drive in capturing them. This starts with an opportunity-seeking outlook and usually involves risk-taking and the adoption of new ideas, new approaches, new products and services, even new customers in new locations. This is predominately a mindset challenge and is directly linked to the leadership and relationship quality factors described earlier. There is also a transition that needs to be made from sticking to core competencies alone to capturing and adjusting to new opportunities that involve doing things very differently. This is a sizable challenge for many SCs

ix Ecosystem awareness – What is changing, what is the likely impact, how should we respond, how might new opportunities be incorporated and realised? This is about awareness of and understanding of what is happening in the external and internal business environment and in particular in the identification of external and internal situations or events that will impact the SC x Competent management of risk. Unfortunately, many SCs and companies within SCs manage risk very poorly. Typical examples of such poor management include over-supply and stock-outs. A detailed risk management tool for the competent and thorough management of risk is presented in Chapter 3

Prima facie, SCE as described earlier is a sizable undertaking. How might it be achieved therefore?

The key premise of this book is that SCE arises from (or is driven by) SC process excellence. The balance of this section, thus, provides an introduction to the four key SC processes listed in Chapter 1, including a description of the level two processes and their key measures. Chapters 3 to 6 then provide a detailed description of each process and its sub-processes, including explanations of practical techniques that can be applied to actually obtain SC process excellence (i.e. 'how to do it' examples).

2.4.1 Key SC process #1 – SC strategy

SC *strategy* is the first, or beginning, process. That is, strategy sets all of SC direction, purpose, key themes, identifies target markets and products or services to be offered to those markets. SC goals and objectives are defined along with key strategies and specific actions to deliver those goals and objectives. SC *strategy* also considers resource requirements and identifies the manner that required resources are to be obtained and developed. The key sub-processes of SC *strategy* are shown in Table 2.1.

Table 2.1 SC Strategy Process Overview

Level One	*Level Two*	*Description*
SC Strategy	Description of 'as-is' SC situation	Includes a comprehensive description of the present SC environment – e.g. customers, suppliers, SC partners, competitors, capabilities, strengths, weaknesses, competitive advantages, social analysis and regulatory compliance conditions
	SC future desired state	Similar items as per previous row only here described as the *desired* future for each
	SC purpose	A clear, concise statement describing the reason(s) the supply chain exists
	SC goals	Specific high-level SC goals and lower-level objectives per future time period
	SC strategies	Overarching strategies and specific actions designed to deliver the goals
	Implementation of SC strategy	A detailed implementation plan describing funding, resources, constraints, processes to be followed and timelines
	Managing SC risks	Identification and rating of SC risks, including probability and consequence scores plus preventative and contingent actions
	SC data and data analysis	Compilation of a formal data specification, data capture, data storage, data cleansing, data analysis, data reporting and results dissemination

Table 2.2 SC Design Process Overview

Level One	*Level Two*	*Description*
SC Design	Design for safety	Comprises a competent *Safety Management System* (SMS) that includes SC safety policy, safety planning (design of products, services, processes, plant, equipment, transportation devices and transport scheduling & routing), risk management, safety implementation, safety measurement & evaluation and management reviews of SC safety
	Design for service	Includes physical SC design, planning and scheduling methods to be adopted, inventory management practices and overall SC operating methodology (e.g. push, pull, flowed-delivery, VMI, Agile)
	Design for quality	Product/service specifications and design, definition of quality and service attributes required, description of process parameters and process control to be used
	Design for sustainability	Modelling of SC features that will deliver target levels of performance on economic, social and environmental outcomes
	Design for integration	Description of how a single end-to-end SC focus is to be achieved, the level of process and personnel integration, role design and a responsible, accountable, consulted, informed (RACI) matrix
	Design for people	Comprehensive description of leadership culture to be followed, administration policies and procedures and desired cultural profile of all SC people
	SC ICT design	SC ICT design including digital SC networks, mobility, Cloud services, software/applications to be used, enabling of process integration, security, information visibility, data management and support
	Design for cost	A description of the ABC costing method to be used, SC cost targets and the methods to be used to achieve those targets
	Design for risk	Failure modes and effects analysis of SC design. SC risk elimination/risk minimisation actions

2.4.2 Key SC process #2 – SC design

SC *design* comprises the translation of SC strategy into an overall 'blueprint' that will facilitate delivery of the SC strategy. This translation is affected via a number of key design elements as shown in Table 2.2.

2.4.3 Key SC process #3 – SC execution

SC *execution* concerns the SC mission critical function of reliable delivery of the customers' value propositions. A value proposition, as the term implies, is the SC's value 'promise' to its customers. The value proposition thus represents the value offered to the customer.

Customers usually assess such value in terms of product and/or service attractiveness, order cycle time and delivery timeliness, SC responsiveness, flexible order lot size, suitable quality and cost. Customers will and do, of course, assiduously compare the offered value against the actually delivered value (i.e. 'Did my supplier actually fulfil their promise to me?')

Factors requiring careful management in order to assure effective SC *execution* therefore are outlined in Table 2.3.

Table 2.3 SC Execution Process Overview

Level One	*Level Two*	*Description*
SC Execution	New Products/services to market process and performance	Description of the process to be followed for new product/service planning, plan deployment, process planning and production planning
	Customer service	Detailed customer offer, customer contracts & service level agreements, product/service catalogues, order capture and entry process, order management process, during and after sales service, measurement and continuous improvement processes
	Order fulfilment	Full details of planning, sourcing, making and delivering processes
	Managing returns	Description of returns notification, return routing, receival, sorting, processing, costs, reporting and continuous improvement
	Enabling suppliers	Categorisation of the supplier base, determination of make/buy/outsource decisions, supplier appraisal processes, supply base rationalisation, supplier development actions, supplier involvement in new products/services, supplier relationships, managing supplier performance
	Collaboration, cooperation, integration	Definition of how SC relationships are to be managed, the goals, strategies and processes to be followed, SC information sharing practices, partner performance measurement, roles, interactions & rewards
	SC execution support systems	Execution of ICT plan from SC design process, including required infrastructure, software/applications, e-Commerce methodology and practice, SC process integration, project management of ICT projects

2.4.4 Key SC process #4 – SC people

In the author's direct experience with real-life operating SCs, the SC *people* process is usually the one most overlooked.

This is perplexing because the author also knows that this process is *the most important* of the four in terms of quality of outcomes and sustainability of performance.

The reasons for such discounted coverage of the people process whilst numerous are common around:

- i *Effective* people management as a topic is underdeveloped; few understand it exhaustively, and even fewer are aware of the full range of issues that need to be managed and therefore how to go about managing them
- ii Until recent times, robust people management training has been missing in most university technical undergraduate degrees
- iii A male-influenced mindset that considers the 'soft' people issues almost not worth worrying about
- iv People have typically been 'looked after' by the Human Resources function within organisations. As such, people management has mostly (but not exclusively) been

Table 2.4 SC People Process Overview

Level One	*Level Two*	*Description*
SC People	SC leadership	Setting direction, leveraging the potential of all resources and ensuring leader technical and social suitability
	People managerial system	Compliance with 'requisite-organisation' methodology, effective management of all SC people costs
	Organisational culture	What are the principles, values and behaviours to be observed by all SC people? What are the required mindsets, alignment, diligence, motivation, work ethic, teamwork, ownership, responsibility, accountability and levels of respect that need to be lived? What is the process for the development and enactment of team charters? How do SC leaders ensure that such a desired culture is actualised?
	Knowledge building	How do SC leaders ensure that SC people are open and inquisitive, actively share ideas, are collaborative learners and active advancers of knowledge via networks, user groups and communities of practice?

about wages, salaries, bonuses and incentives, leave, entitlements, administrative protocols, annual performance reviews, industrial relations and reaction to union demands

The key sub-processes of this most important SC process are show in Table 2.4.

2.5 Proposed theory of supply chain management

In considering an underlying SCM theory, it is important to start with the goals of SCM and how such goals are achieved in practice. In other words, what is SCM really trying to achieve and how does it go about achieving it?

The *goals of SCM*, thus, can be summarised as:

i The delivery of value to customers via the fulfilment of customer defined expectations. In turn, this leads to market share and margin assurance
ii The reduction of all SC costs. This includes the elimination of all forms of waste and minimisation of all inventory costs. In turn, this leads to improved profitability and cash flow
iii The assurance of SC sustainability. This includes economic performance, of course, but also minimisation of the SCs environmental footprint and the fulfilment of formal and informal social 'contracts'

The *tactics* typically used to achieve the SCM goals can be described as:

i The development and delivery to market of attractive products and/or services
ii An end-to-end SC focus and mindset (from raw material suppliers, service providers to end customers)
iii Material, product and service flow management – Includes management of key SC processes such as forecasting, planning and scheduling, modelling, effective constraint management, sourcing, inventory management, capture and launching of

orders, conversion (making), order status reporting, distribution, delivery and delivery confirmation

iv SC asset management – All assets used in the procurement and storage of materials, material conversion, service provision, material transportation, warehousing and delivery, ICT assets and, last but not least, all SC personnel

v Information flow management – The on-line and real-time availability and visibility of key SC information to all those along the SC who need it

vi SC relationship building and maintenance – Engaging and collaborating with all key SC partners towards the delivery of common goals (the previous SC goals). Activities include education and training, coaching, guiding, development of common measures and targets, specific problem solving and improvement projects, auditing, core process redesign, sharing of resources, information sharing and co-location of resources

vii SC process integration – The end-to-end integration of key SC processes (e.g. planning, sourcing, making, delivering and returning processes) so as to improve the performance and sustainability of the overall chain

A theory is a system of rules, procedures and assumptions used to produce a result. A theory can also be expressed as a set of hypotheses related by logical or mathematical argument to explain a wide variety of connected phenomena in general terms.

It is proposed that the rules, procedures and assumptions of SCM are contained and described in the four-process framework that is the skeleton of this book. This framework expands into hundreds of sub-topics, each of them defining a single aspect of SCM that must be managed if the overall goals of SCM (defined earlier) are to be achieved.

Utilising the four-process framework, the basic hypotheses of SCM can be developed as:

i Improvements to the levels 4, 5 and 6 factors defined in the SCM four-process framework will result in improved outcomes for each of the four level one processes

ii Improvements to the four level one processes' outcomes will lead to SC *technical* and *cultural* excellence

iii SC *technical* and *cultural* excellence will result in improved SC outcomes, in particular:

   a The development of attractive products and/or services
   b Delivery of customer orders in-full, on-time and error free (DIFOTEF)
   c Reduced order cycle times (order lead time)
   d Responsiveness to changing customer demands and/or changing supply conditions and/or changing conversion conditions (i.e. SC resilience)
   e Lower days of inventory in the entire SC
   f Reduced total SC costs (Activity-Based Costs [ABC])
   g Improved social responsibility results

iv Improvements to key SC outcomes will result in improved organisational economic performance, improved sustainability performance and thus support stakeholder value-add and organisational survival and growth

The proposed theory of SCM can be visualised as shown in Figure 2.6. The relationships shown in Figure 2.6 basically describe the logic of this book: the desired outcomes of SCM and the key drivers of such outcomes.

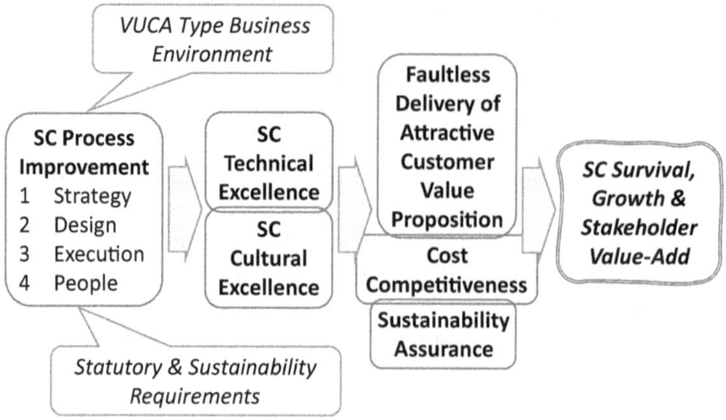

Figure 2.6 Proposed SCM Theory Framework

2.6 Chapter 2 case study – dealing effectively with supply chain challenges

The case study for this chapter concerns the ABC company, an Indian automobile parts and farm machinery manufacturer. The case study is based on a research paper written by Siva Kumar and Ramesh Anbanandam (2020).

The authors set out to test the usefulness of a so-called SAP-LAP framework (first proposed by Sushil (2000, 2001), in helping companies such as ABC to develop a SC Resilience (SCRes) building processes in order to improve their level of SCRes in the context of volatility and SC disruptions. The main elements of the SAP-LAP framework are shown in Figure 2.7. Other findings, of course, can be used to complement such a framework. For example, Dubey et al. (2019) established a significant relationship between the use of data analytics and organisational flexibility on SCRes levels. Such findings can therefore be included in any SC's SAP-LAP framework content.

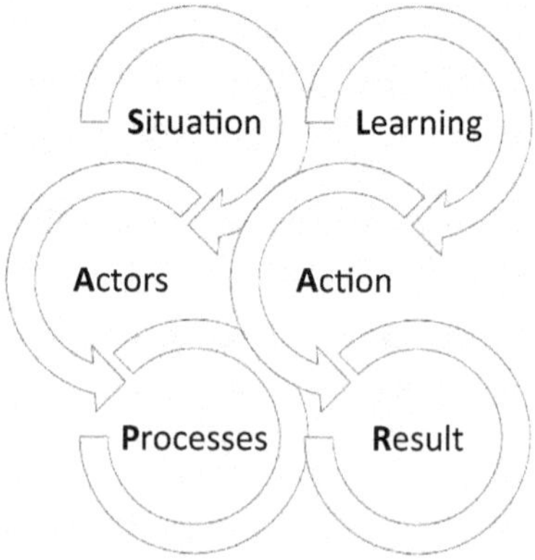

Figure 2.7 Main Elements of the SAP-LAP Framework

SCRes is defined variously in the literature; however, for this case study summary, SCRes is defined as: *The ability of a SC to withstand disruptions to any or all of its expected supply inputs, conversion operations and customer demand levels.* Definitions of what would constitute 'withstanding disruptions' varies from SC to SC; that is, some SCs might require that SC disruptions as high as ± 50% must be able to be managed, whereas other SC may set the limit at ± 25%. The higher the limit, the more the SC would be resilient to any such unexpected shocks. And of course, the higher the limit, the more work any given SC must do in order to effectively and efficiently manage such events.

Three research questions were posed by Kumar and Anbanandam (2020):

- i How can organisations build resilience into SCs and improve it continuously?
- ii What are the key processes that infuse resilience into SCs?
- iii What are the key measures or actions that enhance an organisation's ability to manage SC disruptions effectively and efficiently?

In their literature review search, Kumar and Anbanandam (2020) found no overall and inclusive theory. That is, they found many articles dealing with the quantitative side of building SCRes; however, they identified a gap on the qualitative side (i.e. little research work has been conducted on the study of values, principles, behaviours, trustworthiness and cooperation when it comes to SCRes). They pointed this out as a deficiency because the building of SCRes is a process that takes a long period of joint effort and collaboration. It is for this reason they were drawn to explore the SAP-LAP framework in order to assist SC professionals in improving their SC's resilience.

After a series of ABC company plant visits and interviews/consultations with managers and examination of company reports, Kumar and Anbanandam (2020) described the following results:

- i Limited awareness of SC risks existed
- ii SC risk management culture was underdeveloped
- iii Moderate information sharing between SC partners took place
- iv Limited knowledge of suppliers' and other SC partners' business continuity plans existed
- v No formal business continuity plan validation process existed
- vi The level of genuine SC partner collaboration was found to be low

In response, and to assist the ABC company lift its level of SCRes, Kumar and Anbanandam (2020) proposed the following:

- i That use be made of the LAP side of the SAP-LAP framework developed, such that company managers can work to build ABC's level of SCRes
- ii That genuine effort should be made to include all relevant SC partners in the work of building SCRes
- iii That discovery and development of ways to motivate SC partners and senior leaders be undertaken in order for them to better support an active SCRes building program

Taking all of the information and learning from the case and elsewhere, a '*Supply Chain Resilience Builder*' is now proposed and is shown in Figure 2.8.

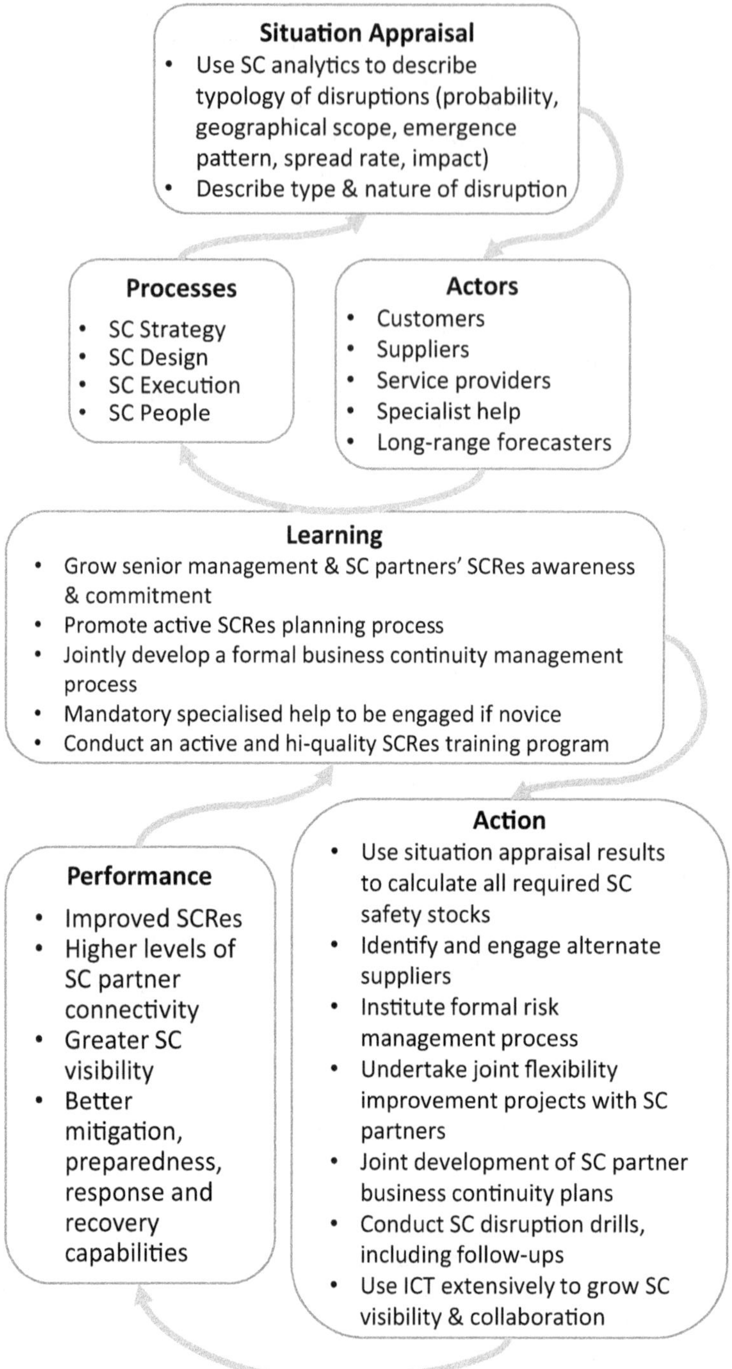

Figure 2.8 Supply Chain Resilience Builder

2.7 Review questions

i What is meant by a VUCA type business environment?
ii What are the seven challenges that a VUCA type environment presents to SCM?
iii What are the four response essentials in meeting the VUCA driven SC challenges?
iv What are the main differences between SC *technical* excellence and SC *cultural* excellence?
v Can a SC survive long term on either SC *technical* excellence or SC *cultural* excellence alone?
vi What are the four key SC processes?
vii Which of the four key SC processes are the more important one(s) in your direct (or indirect) experience? Why so?
viii What is the logic of the SC theory framework proposed in this chapter?

2.8 Assignment topics

(Each to be 3,000 words minimum or about three A4 pages single spaced. Tables and figures are not to be included in the word count.)

i From the results of your own research, describe the drivers of *Volatility, Uncertainty, Complexity* and *Ambiguity* (VUCA)
ii Describe the four VUCA response essentials in some detail. From your research, what obstacles might a SC professional encounter in attempting to apply each response essential?
iii Describe in some detail the essential differences between SC *technical* excellence and SC *cultural* excellence? Why is it that these two factors are a duet with respect to SC performance?
iv SCE is a major undertaking for any SC. Describe the four SC processes that can be worked on and improved in order to actually achieve SCE
v Provide a critique of the proposed SC resilience builder presented in Figure 2.8. In your view, would such a process be helpful or unhelpful in building SC resilience? Provide evidence to support your position

2.9 Chapter 2 references

Collins, J. (2001). *Good to Great*. Sydney: Random House.
Dubey, R., Gunasekaran, A., Childe, S. J., Fosso Wamba, S., Roubaud, D., Foropon, C. (2019). Empirical Investigation of Data Analytics Capability and Organizational Flexibility as Complements to Supply Chain Resilience. *International Journal of Production Research*. https://doi.org/10.1080/00207543.2019.1582820.
Kim, W. C., Mauborgne, R. (2005). *Blue Ocean Strategy*. Boston, MA: Harvard Business School Press.
Ohno, T. (1988). *Toyota Production System*. Boca Raton, FL: Taylor & Francis.
Siva Kumar, P., Anbanandam, R. (2020). Theory Building on Supply Chain Resilience. *Global Journal of Flexible Systems Management*, 21(2), pp. 113–133.
Sushil. (2000). SAP – LAP Models of Inquiry. *Management Decision*, 38(5), pp. 347–353.
Sushil. (2001). SAP-LAP Framework. *Global Journal of Flexible Systems Management*, 2(1), pp. 51–56.

3 Setting an appropriate supply chain *strategy*

3.1 What you will learn in this chapter

This chapter starts with developing an understanding of three useful tools that, if applied competently, can be used to provide a professional assessment of any given SC's current and likely future *situation*.

Next, coverage of the processes that can be used to develop a SC's *vision, purpose, goals, strategies, implementation plans* and to conduct a thorough *risk analysis*.

SC data, data analysis processes and metrics are then presented, followed by the deeper processes used to conduct a robust situation analysis.

The elements of a *sound SC strategy* are then discussed along with how to go about testing for strategic alignment and testing the efficacy of any chosen SC strategy.

Next, SC strategy implementation *pitfalls and obstacles* are described in two main groups: *technically* based obstacles and *culturally* based obstacles.

Finally, a case study of a real-life SC *strategy option testing* project for production of the sustainable fuel bioethanol is described.

3.2 *Strategy* sub-processes and metrics

Level two and level three SC *strategy* processes are introduced in this section. Importantly, the metrics used to assess if the level three processes are performing at a level required to underpin SC *strategy process excellence* are also defined.

In setting a competitive SC strategy, it is important to at first understand the reality of the situation confronting the SC. To do this, it is necessary to conduct an analysis of the business environmental conditions that exist. Some refer to such an analysis as an 'as-is' situation appraisal.

Fortunately, several tools exist to help SC leaders and SC professionals with such an analysis. Three such tools are presented next: a political, economic, social, technological, legal and environmental (PESTLE) analysis; a strengths, weaknesses, opportunities and threats (SWOT) analysis and a vision elements analysis (VEA). Table 3.1 summarises the three approaches and their respective metrics.

Following completion of the situation analysis and discussion about its veracity and implications, it is time to jointly develop a lucid description of the future desired state of the SC. Some refer to this as visioning.

Once the *vision* is set and agreed, the SC *purpose* needs to be agreed and shared, the desired and specific SC *goals* set, the *strategies* to achieve those goals defined, the *implementation plan* for those strategies developed and lastly, a full SC *risk analysis* needs to be undertaken and completed.

Table 3.1 'As-Is' Situation Analysis Tools

Level Two	*Level Three*	*Metrics Used for Level Three*
Description of 'as-is' situation	Political, Economic, Social, Technological, Legal, Environmental (PESTLE) analysis	A professionally prepared PESTLE analysis exists and is documented Target rating is specified for each PESTLE element Derived implications of PESTLE analysis exist
	Strengths, Weaknesses, Opportunities, Threats (SWOT) analysis	A professionally prepared SWOT analysis exists Target rating specified for each element Derived implications of SWOT analysis exist
	Analysis of 'as-is' situation for each vision element	Each vision element has been professionally rated and documented Derived implications of vision elements analysis exist

Table 3.2 Level Two and Three Processes for Setting of SC Vision, Purpose, Goals, Strategies, Implementation Plans and Risk Analysis

Level Two	*Level Three*	*Metrics Used for Level Three*
Description of SC future desired state (Vision)	Identification of the elements of the SC vision	SC vision element standard
	Description of the desired state for each element	Target values per vision element
	Overarching supply chain 'vision' statement	Number of words, clarity, ownership level
Supply Chain Purpose	Overarching 'mission' statement	Number of words, clarity, ownership level
SC Goals	Overarching goal per vision element	Index of goal's ability to deliver the vision element
SC Strategies	Overarching strategies designed to deliver goals	Index of strategy's ability to deliver the goal
	Statement of SC strategy alignment to business strategy	For each strategy, an index of alignment to overall business strategy
Implementation of SC Strategy	Detailed implementation plan	Index of plan's ability to deliver strategies and goals
SC Risks	Identification & rating of each major SC risk	Risk rating & categorisation

The level two and level three processes involved with this work and the metrics used for the level three processes are shown in Table 3.2.

Due to the reality that present-day SCs generate and store enormous amounts of data, it is important to include SC data and data analysis as part of any comprehensive SC strategy. Table 3.3 lists the relevant level three SC data processes and key metrics that can be used to assess the level three process performance.

Before describing the three 'as-is' situation analysis tools in more detail, it is important to note that, in order to conduct such analyses competently, it is necessary to *conduct rigorous research and analysis* that usually involves considerable time and effort. For such analyses to be useful, they cannot be superficial or someone's 'best guess'. Meaning, they cannot be competently compiled by a group of leaders with no pre-work, sitting around

Table 3.3 Level Three SC Data and Data Analysis Processes and Metrics

Level Two	*Level Three*	*Metrics Used for Level Three*
SC Data and Data Analysis	Identify and define supply chain data requirements	Detailed data specification exists
	Data collection and storage procedure	Approved data collection and storage procedure exists Data collection and storage cost targets
	Data inspection and cleansing procedure	Data inspection and cleansing procedure exists Total data errors/total data elements * 100 goal Inspection and cleansing duration and cost targets
	Analysis of cleansed data procedure	Analysis of cleansed data procedure exists, including identification of the data analysis techniques, how they work and how they are applied Index of analysis quality target Analysis duration and cost targets
	Results recording, analysis report compilation and reports dissemination procedure	Results recording, analysis report compilation and reports dissemination procedure exists Targets exist for duration and cost of report compilation

Table 3.4 Levels Four, Five and Six of the 'Political' Element of the PESTLE Analysis

Level Four	*Level Five*	*Level Six*
Political	Stability	Tenure of political leaders Tenure of elected members Tenure of ruling parties War and conflicts
	Attitude Towards Business	Tax policies Levels of 'red tape' Support for growth of existing businesses Support for start-up of new businesses Commitment to spending on sensible infrastructure Encourage and support industry specific education
	Attitude Towards Citizens	Health policies Housing policies Education policies Welfare policies
	Governance Quality	Governance processes Effective political control Level of corruption
	Quality of Political Leaders	Competence Experience Responsible Respectable & respected

in a conference room for 45 minutes, conducting a PESTLE analysis, for example, and convincing themselves that they have conducted an in-depth appraisal. To understand this statement more completely, the three situation analysis tools will now be considered more deeply.

The first such tool is the PESTLE analysis, and the level four, five and six processes for the 'Political' element are illustrated in Table 3.4.

Table 3.5 Levels Four, Five and Six of the 'Economic' Element of the PESTLE Analysis

Level Four	*Level Five*	*Level Six*
Economic	% GDP Growth	
	Interest Rates (%)	Savings Borrowings
	Budget Deficit as % of GDP	
	NET Public Debt as % of GDP	
	Trade Balance (US$B)	
	% Inflation	
	Exchange Rate to Major Currencies	USD, EUR, CNY, JPY, GBP, RUB, CHF, CAD, AUD
	Cost of Labour ($/hour)	White collar Blue collar – skilled Blue collar – unskilled
	Productivity of Labour (output of goods or services divided by labour hours to produce same)	White collar Blue collar – skilled Blue collar – unskilled
	Transport Availability (Route by product type capacities per time period)	Air Sea/canal/river Rail Truck Pipeline
	Transport Costs (Route by product type $/unit)	Air Sea/canal/river Rail Truck Pipeline
	Finance availability (credit availability index)	Banks Government Investors
	Competitors	Number and spread Competitor strength and energy

Next are the level four, five and six elements for the 'Economic' element of the PESTLE analysis as shown in Table 3.5.

The third element of the PESTLE analysis is 'Social', and its level four, five and six elements are illustrated in Table 3.6.

The fourth element of the PESTLE analysis is 'Technological', and its level four, five and six elements are illustrated in Table 3.7.

The fifth element of the PESTLE analysis is 'Legal', and its level four, five and six elements are illustrated in Table 3.8.

The last element of the PESTLE analysis is 'Environmental', and its level four, five and six elements are illustrated in Table 3.9.

The second situation analysis tool is the SWOT analysis. The first two elements, Strengths and Weaknesses, are more SC internally focused and the last two, Opportunities and Threats, are more SC externally focused.

Starting with the 'Strengths' element, its level four, five and six elements are illustrated in Table 3.10.

Next is the 'Weaknesses' element, and its level four, five and six processes are illustrated as shown in Table 3.11.

Table 3.6 Levels Four, Five and Six of the 'Social' Element of the PESTLE Analysis

Level Four	*Level Five*	*Level Six*
Social	Community Attitude Towards Business	Pro-business Vs CAVE people (Citizens Against Virtually Everything)
	Level of Community Engagement with Business	Regular and willing Vs infrequent and unwilling
	Average Wealth of Community	Total assets/capita Total cash/capita
	Wealth Distribution of Community	x% of population holds y% of wealth (e.g. 20% of population holds 80% of wealth Vs 80% of population holds 80% of wealth)
	Age Distribution of Community	Shape of curve (e.g. normal bell curve or skewed left or skewed right or bi-modal)
	Political Affiliation of Community	Conservative (pro-business, free-trade & open market) Vs Socialist (State ownership & control, wealth levelling)
	Ethnic/Religious Factors	Nil Vs highly prevalent and disruptive
	Attitude Towards Employment	Willing and eager to work Vs 4th generation welfare recipients
	Lifestyle Trends	Static Vs strong desire and energy for lifestyle change
	Customer Consumption Patterns	Stable Vs highly variable Vs cyclical (e.g. seasonal)
	Consumer Brand Recognition	Very low (generic brands only available) to very high (high brand loyalty apparent)
	Attitude Towards Well-Being of Self	Health and safety (physical and mental) Quality of life and living standard Wealth and retirement provisions Education, skills and competencies Career progression

Table 3.7 Levels Four, Five and Six of the 'Technological' Element of the PESTLE Analysis

Level Four	*Level Five*	*Level Six*
Technological	Level of & consistency of spending on R&D	Basic research Applied research Development
	Quality Access to Internet	Availability Speed Reliability
	Availability of leading process technologies	Materials processing Information processing Human processing (e.g. health, education, transport, shopping, banking, communications)
	Availability of leading product technologies	Proprietary Non-proprietary
	Availability of leading transport & distribution technologies	Air Water Land Intra- and intercontinental tubes
	Degree of Automation used in Industries	Mechanisation Process computers Robotics
	Rate of technological change	Rate of change of outputs/inputs ratio

Table 3.8 Levels Four, Five and Six of the 'Legal' Element of the PESTLE Analysis

Level Four	*Level Five*	*Level Six*
Legal	Financial reporting and taxation	Registered office requirements
		Business name and number
		Financial records and financial reports
		Taxes and fees
	Environmental Legislation	International Environmental Law
		National Acts and Statutes
		State Acts and Statutes
	Occupational Health & Safety (OH&S) Legislation	International Labour Organisation (ILO) Occupational Safety and Health Standards
		National Health and Safety Acts (e.g. Safe Work Australia Act, 2008)
		State OH&S Acts and Statutes
	Discrimination Laws (sometimes referred to as anti- or non-discrimination laws)	International Human Rights Legal Framework
		National Discrimination and Equal Opportunity Acts
		State Discrimination and Equal Opportunity Acts
	Consumer Laws	UN Guidelines for Consumer Protection
		National Consumer Laws
		State Fair Trading Laws
	Competition Laws	WTO Trade Rules
		National Competition Laws
		State Competition Laws
	Labour/Employment Laws	ILO International Labour Standards
		National Labour Laws
		State Labour Laws
	Trade Restrictions and/or Tariffs	Tariffs
		Quotas
		Trade Licences
		Local Content Requirements
	Sovereign Risk	Budget Deficit
		National Debt
		Trustworthiness of political leaders
		Robustness and fairness of Country/State legal system

Table 3.9 Levels Four, Five and Six of the 'Environmental' Element of the PESTLE Analysis

Level Four	*Level Five*	*Level Six*
Environmental	Compliance	Total cost of compliance (capital, personnel, time)
		Cost of non-compliance (fines, loss of operating licence, lawsuits)
	Risk	Number of possible adverse environmental 'events'
		Likelihood of adverse environmental 'event'
		Consequence of an adverse environmental 'event'
	Community Attitude Towards Environmental Issues	Number of formally organised environmental groups
		Power and influence of environmental groups
		Attitude of environmental groups towards business
		Likelihood of new environmental groups forming
	Recycling Requirements	National Waste Policies
		State Waste Regulations
	Environmental Taxes (e.g. Carbon Tax)	Pollution taxes (air, water, soil, noise)
		Revenue-Based Taxes (e.g. congestion tax, fuel excise tax)
		Incentivised Taxes (e.g. renewable energy subsidies)

(*Continued*)

Table 3.9 (Continued)

Level Four	*Level Five*	*Level Six*
	Climate and Weather	Temperature (minimum, maximum, average) Rainfall (annual total and monthly pattern, plus patterns of floods and droughts) Wind, storms, cyclones, tornadoes
	Impact of Climate Change	Temperature impacts actually experienced Rainfall impacts actually experienced Adverse weather events actually experienced and impact Rising sea levels actually experienced and impact Agricultural effects actually experienced Health effects actually experienced Infrastructure impacts actually experienced

Table 3.10 Levels Four, Five and Six of the 'Strengths' Element of the SWOT Analysis

Level Four	*Level Five*	*Level Six*
Strengths	People (competence, experience, motivation, innovativeness, responsibility)	Leaders White collar Blue collar – skilled Blue collar – unskilled
	Processes (modernity, patents, technology)	Strategy Design Execution People
	Products/Services	Attractiveness Fit for purpose rating Quality Affordability Availability Order cycle time Delivery reliability
	Plant and Equipment	Age Efficiency Capability Capacity Reliability Flexibility
	Pace	Cycle times Responsiveness Adaptability Agility
	ICT Systems	Customer facing Supply chain partner facing Employee facing Structure, capacity and costs Use of Big Data and SC analytics
	Real competitive advantages	Effective research and development Attractiveness of product/service offer People competence and commitment Raw materials quality and availability Order cycle time leadership Genuine low-cost producer Differentiated supplier
	Economic health	Cash position Shareholder Value Add (SVA) Financing and leverage

Table 3.11 Levels Four, Five and Six of the 'Weaknesses' Element of the SWOT Analysis

Level Four	*Level Five*	*Level Six*
Weaknesses	People (competence, experience, motivation, innovativeness, responsibility)	Leaders White collar Blue collar – skilled Blue collar – unskilled
	Processes (modernity, patents, technology)	Strategy Design Execution People
	Products/Services	Attractiveness Fit for purpose rating Quality Affordability Availability Order cycle time Delivery reliability
	Plant and Equipment	Age Efficiency Capability Capacity Reliability Flexibility
	Pace	Cycle times Responsiveness Adaptability Agility
	ICT Systems	Customer facing Supply Chain partner facing Employee facing Structure, capacity, flexibility and costs Use of Big Data and analytics
	Competitive disadvantages	Research and Development Level of attractiveness of product/service offer People competence and commitment Raw materials quality and availability Comparative order cycle time position Comparative cost position Differentiated supplier(s)
	Economic health	Cash position Shareholder Value Add (SVA) Financing and leverage

The third SWOT analysis element is 'Opportunities', and its level four, five and six elements are illustrated as shown in Table 3.12.

The last SWOT analysis element is that of 'Threats', and its level four, five and six elements are illustrated in Table 3.13.

The third and last situation analysis tool to be covered here is that of appraising each of the SC's vision elements. The level four vision elements and their corresponding level five elements are shown in Table 3.14 and 3.15.

As can be seen from Tables 3.4 to 3.15, conducting a sound assessment of the deepest level of each analysis tool is a sizeable task and not one to be taken lightly or under-resourced. This is really a job for a well-qualified and competent SC analyst.

Table 3.12 Levels Four, Five and Six of the 'Opportunities' Element of the SWOT Analysis

Level Four	*Level Five*	*Level Six*
Opportunities	Community perception	Special interest groups
		Media coverage
		Community engagement
	Markets	International
		National
		Regional
	Customers	New customers
		Retention of existing customers
		Growth of existing customers
		Trends in customer behaviour/preferences
	Supply chain collaborators	Suppliers
		Service providers
		Technology providers
	Competitors	Existing Competitors
		New Entrants
	Technology	Products/Services
		ICT Systems
		Data science
	Government subsidies/incentives	Capital investment incentives
		Jobs creation incentives
		Renewable energy subsidies
	Government legislation, regulations	Consumer laws
		Trade and competition
		Occupational health and safety
		Environmental laws and regulations
		Labour relations
		Discrimination and equal opportunity
		Foreign investment
		Taxation, fees and licences
	Bankers/Financiers	Access to funds
		Cost of funds
		Transaction ease enabled by technology
		Risk reduction from stable funders

Table 3.13 Levels Four, Five and Six of the 'Threats' Element of the SWOT Analysis

Level Four	*Level Five*	*Level Six*
Threats	Community perception	Activist type special interest groups
		Negative media coverage
		Community disengagement
	Markets	International downturn
		National recession
		Volatile regional demand
	Customers	Likelihood of attracting new customers is low
		Loss of existing customers
		Weak growth of existing customers
		Trends in customer behaviour/preferences
	Supply chain partners	Suppliers (e.g. material shortages and/or bankruptcies)
		Service providers (e.g. overcharging, reneging on contracts)
		Technology providers (e.g. sign up with competitors)
	Competitors	Growing existing competitors' intensity
		Likely new entrants

Level Four	*Level Five*	*Level Six*
	Technology	R&D failures on new products/services
		ICT Systems seen as too costly, too risky
		Data science seen as too complicated
	Government subsidies/ incentives	Capital investment incentives withdrawn
		Jobs creation incentives reduced
		Renewable energy subsidies removed
	Government legislation, regulations	Tougher consumer laws
		Trade and competition rules tightened
		Occupational health and safety (e.g. compliance costs increased by 25%, penalties doubled)
		Environmental requirements heightened
		Labour relations more arduous
		Discrimination and equal opportunity requirements raised
		Foreign investment rules relaxed
		Taxation, fees and licences increased
	Bankers/ Financiers	Capital freeze, access to funds limited
		Higher cost of funds
		Transaction security failures caused by technology
		Risk increase from unstable funders

Table 3.14 The First Five Elements of the Vision Elements Appraisal Analysis

Level Four	*Level Five*
Customers	Number of customers
	Customers' size
	Customers' competitive advantages
	Customers' willingness to collaborate
	Customers' satisfaction with our performance
	Customers' openness to new ideas
	Customers' continuous improvement strength
	Customers' market power
	Customers' buying behaviour & preference trends
Suppliers	Number of suppliers
	Suppliers' size
	Suppliers' competitive advantages
	Suppliers' willingness to collaborate
	Suppliers' openness to new ideas
	Suppliers' continuous improvement strength
	Suppliers' market power
Service Providers (SPs)	SPs' competitive advantages
	SPs' willingness to collaborate
	SPs' openness to new ideas
	SPs' continuous improvement strength
Competitors	Number of competitors
	Competitors' size
	Competitors' competitive advantages
	Competitors' aggressiveness
	Competitors' market power
Community	Level of community engagement and communication
	Level of community initiatives
	Success of community initiatives
	Community's attitude towards our operations

Table 3.15 The Last Five Elements of the Vision Elements Appraisal Analysis

Level Four	*Level Five*
Technical Capabilities	Capability of products, services and brand
	Processes capability
	ICT technologies capability
Competitive Advantages	Products/services/brand
	Processes
	People
	Differentiated customer offer
	Channels to market strength
	Cost competitiveness
	Customer loyalty
People	Numbers of
	Skills/capabilities/competencies
	Motivation/energy
	Turnover rates (churn)
Regulatory compliance	Occupational health and safety laws
	Competition laws
	Environmental laws
	Consumer protection laws
	Employment laws
	Financial reporting rules and regulations
Sustainability	Carbon footprint reduction
	Energy usage reduction
	Waste reduction
	Social responsibility
	Recycling
	Routing optimisation
	Economic health

Table 3.16 A 5-Point Likert Rating Scale for the Appraised Elements

Rating Scale	
1	Very Small/Very Weak/Very Unstable/Very Unattractive
2	Small/Weak/Unstable/Unattractive
3	Average/Average/Average/Average
4	Large/Strong/Stable/Attractive
5	Very Large/Very Strong/Very Stable/Very Attractive

Each of the deepest elements for each tool, can be rated against a 5- or 7-point Likert scale and compared to desired targets for each element. Such a rating scale is shown in Table 3.16.

The results of such a proper analysis can not only be highly educative for SC personnel but can also identify areas that need to be addressed, the urgency with which they need be addressed and also spark innovative ideas that, if approved and implemented, can grow competitive advantage.

3.3 The elements of a sound supply chain *strategy*

The elements of a sound SC strategy consist of a *vision*, a *purpose*, *SC objectives* and *goals*, *specific strategies* to be pursued in order to achieve said goals, an *implementation plan* and a full *risk and response analysis*.

Starting with an appropriate SC *vision*, it is at first necessary to define the vision elements and then to describe the desired future state for each of the vision sub-elements. The vision elements are identified in Tables 3.14 and 3.15. Taking the vision elements one at a time, the sub-elements and future desired state for 'Customers', for an example SC, are shown in Table 3.17.

Table 3.17 Future Desired State for Each 'Customers' Sub-Element (for an Example SC)

SC Vision Elements	*SC Vision Sub-Elements*	*Written Description (an Example)*	*Desired Rating*
Customers	Attractiveness of product/service offer	Our desire is to offer our customers a superior value proposition and to deliver it faultlessly. We want our products and services to be differentiated on the basis of attractiveness, suitability for purpose, quality, durability and competitive order cycle times. Our customer supply chain touch points must include ease of product/service selection, ease of ordering, reliable order status reporting, delivery reliability, and after sales support	5
	Delivery of product/ service offer	Our desire is to deliver our product/service offer with high levels of delivery reliability (98% DIFOTEF) and competitive order cycle times. We want to be responsive to changing customer requirements with sufficient supply chain flexibility and agility to do this	5
	Number of customers	Our desire is to have greater than 1,000 customers none of whom dominate in size or power	4
	Customers' size	Ideally no customer has a size greater than 15% of our total supply	4
	Customers' competitive advantages	We cherish customers with strong brands, attractive products and services, solid reputations and ethical behaviours	5
	Customers' willingness to collaborate	Our desire is to reward customers who enthusiastically embrace collaboration and delivery of joint supply chain goals. Benefits obtained from such collaboration will be shared in proportion to each party's level of investment in the initiative	5
	Customers' satisfaction with our performance	We desire a customer satisfaction rating of 5	5
	Customers' openness to new ideas	We will reward customers displaying an openness to new ideas	4
	Customers' continuous improvement strength	We desire to work actively with customers in utilising our accredited continuous improvement process in order to improve supply chain outcomes	5
	Customers' market power	Our desire is to have no customer who dominates in size and/or market power	3
	Customers' buying behaviour & preference trends	Our desire is to completely understand our customers' buying behaviours and preferences and that through collaborative efforts such behaviours and preferences are promptly responded to	4

(*Continued*)

Table 3.17 (Continued)

SC Vision Elements	*SC Vision Sub-Elements*	*Written Description (an Example)*	*Desired Rating*
	Our pricing to customers	Pricing of our products and services will vary depending on the level of differentiation we are able to bring to market for each product and service. Overall, we desire to achieve a positive Shareholder Value Add result with a Return on Capital margin of 2%. Such a result will be achieved by tight management of costs and setting prices accordingly. If the market will not bear the price that such an approach determines, then we must either reduce the total supply chain cost to market for the products or services in question or no longer supply them	4

Next, the sub-elements and future desired state for 'Suppliers', for an example SC, are shown in Table 3.18.

The sub-elements and future desired state for 'Service Providers', for an example SC, are shown in Table 3.19.

The sub-elements and future desired state for 'Competitors and Community', for an example SC, are shown in Table 3.20.

The sub-elements and future desired state for 'Technical Capabilities', for an example SC, are shown in Table 3.21.

The sub-elements and future desired state for 'Competitive Advantages', for an example SC, are shown in Table 3.22.

Lastly, the sub-elements and future desired state for 'People, Regulatory Compliance and Sustainability', for an example SC, are shown in Table 3.23.

After describing the future desired state for each of the vision elements, some organisations prefer to capture the essence of such a description in a concise *vision* statement. An example of a *vision* statement for the aluminium smelter AlySmelt could be as simple as:

> *We want to be a highly valued and respected partner in a very successful aluminium supply chain.*

Whilst this statement does not really say much, it alludes to all of the individual future desired state descriptions.

In a similar way, some organisations prefer to also produce a concise *purpose* statement. So again, for AlySmelt:

> *We exist to completely satisfy our aluminium supply chain customers profitably.*

Following the future desired state descriptions and the crafting of relevant *vision* and *mission* statements, it is important to specify the goals sought for each vision element that will, when achieved, deliver the desired vision element result. The goals for the same example SC are shown in Table 3.24.

Following goal setting, it is time to craft the specific strategies to be applied to actually achieve the set goals. Again, for the same example SC, the overarching SC strategies and

Table 3.18 Future Desired State for Each 'Suppliers' Sub-Element (for an Example SC)

SC Vision Elements	*SC Vision Sub-Elements*	*Written Description (an Example)*	*Desired Rating*
Suppliers	Attractiveness of product offer	We reward suppliers who offer attractive products on competitive order cycle times	5
	Reliability of delivery of product offer	We reward suppliers who reliably deliver their product offers. Our expectation is 98% DIFOTEF	5
	Number of suppliers	Our desire is to have a small number (≤5) of reputable and reliable suppliers for each of our major stock keeping units (SKU) material purchases	4
	Suppliers' size	Our desire is to have no supplier dominate in size or power	3
	Suppliers' competitive advantages	We reward suppliers achieving genuine competitive advantages. Our preference is for suppliers who reliably supply quality and fit for purpose products. Supply chain competencies, solid reputation and ethical behaviour are also key determinates in our buying choices	5
	Suppliers' willingness to collaborate	Our desire is to reward suppliers who enthusiastically embrace collaboration and delivery of joint supply chain goals. As for customers, benefits obtained from such supply chain collaboration will be shared in proportion to each party's level of investment in the initiative	5
	Suppliers' openness to new ideas	We reward suppliers displaying an openness to new ideas	5
	Suppliers' continuous improvement strength	We desire to work actively with suppliers in utilising our accredited continuous improvement process in order to improve supply chain outcomes	5
	Suppliers' market power	Our strong desire is to have no supplier who dominates in size and/or market power	3
	Suppliers pricing	Our purchasing requirements will be segmented, and our pricing desire is that the price we pay is in the bottom quartile for each segment. This may vary depending on the strategic importance of the supplies and the suppliers involved	4

Table 3.19 Future Desired State for Each 'Service Providers' Sub-Element (for an Example SC)

SC Vision Elements	*SC Vision Sub-Elements*	*Written Description (an Example)*	*Desired Rating*
Service Providers (SPs)	Attractiveness of service offer	We cherish service providers who offer attractive services that are affordable	5
	Reliability of delivery of service offer	We reward service providers who reliably deliver their service offer on-time, on-budget and to-promised functionality	5
	SPs' competitive advantages	Attributes such as supply chain competencies, technological edge, ease of start-up, ease of use, reliability of service, heightened customer experiences, professionalism and ethical behaviour are all very much valued by us	5
	SPs' willingness to collaborate	Our desire is to reward service providers who enthusiastically embrace collaboration and delivery of joint goals	5
	SPs' openness to new ideas	We reward service providers displaying an openness to new ideas	5
	SPs' continuous improvement strength	We desire to work actively with service providers in utilising our accredited continuous improvement process to improve supply chain outcomes	5
	SPs' pricing	Our desire is to negotiate pricing for any new service projects on a turn-key basis. For ongoing services, we prefer to pay for services used in accord with an up-front agreed pricing formula	4

Table 3.20 Future Desired State for Each 'Competitors and Community' Sub-Element (for an Example SC)

SC Vision Elements	*SC Vision Sub-Elements*	*Written Description (an Example)*	*Desired Rating*
Competitors	Number of competitors	Our desire is to not compete in markets where many hungry and hostile competitors exist. This means our product and service offers will be differentiated such that customers are prepared to pay a premium for our superior offers	2
	Competitors' size	This is largely beyond our control; however, our preference is to have a balance of competitor sizes	1
	Competitors' competitive advantages	Our desire is to better our competitors' competitive advantages through enhanced supply chain processes and improvement techniques	1
	Competitors' aggressiveness	Our desire is to have sufficient supply chain fitness to be able to legally manage any competitor aggressiveness	1
	Competitors' market power	Our preference is to avoid head-to-head market battles with competitors who are all powerful. So we will strive instead for alliances or choose another market or industry area	2

SC Vision Elements	*SC Vision Sub-Elements*	*Written Description (an Example)*	*Desired Rating*
Community	Level of community engagement and communication	Our desire is to maintain a level of community engagement and open communications such that we build trustful relationships with our SC community	5
	Level of community initiatives	We wish to embark on consultations with our SC community that lead to agreement on the actioning of shared community initiatives	4
	Success of community initiatives	Our desire is that our community initiatives be successful through careful planning, community collaboration and careful execution	5
	Community's attitude towards our operations	We very much want to be in the position where our community considers us to be valued community members	5

Table 3.21 Future Desired State for Each 'Technical Capabilities' Sub-Element (for an Example SC)

SC Vision Elements	*SC Vision Sub-Elements*	*Written Description (an Example)*	*Desired Rating*
Technical Capabilities	Products, services, brand capability	Our supply chain technical capabilities will arise from a combination of superior supply chain people, the development of attractive products and services and the prudent purchase of modern but proven technologies	5
	Processes capability	We have identified the supply chain processes that matter the most to supply chain performance, and we apply reliable improvement methodologies to those processes in order to achieve supply chain excellence	5
	ICT technologies capability	Supply chain performance is very much enabled by the competent design and application of ICT systems. Our desire then is to very carefully design our requirements and then to select ICT hardware and application solutions that match those designs. Integral to our approach will be the building of a fully digital supply chain, the sensible adoption of Cloud Computing, the active use of Supply Chain Analytics, selected usage of Artificial Intelligence (AI), Big Data and Data Mining, Augmented Reality (AR), Blockchain for traceability, robotics and autonomous transport vehicles	5

Table 3.22 Future Desired State for Each 'Competitive Advantages' Sub-Element (for an Example SC)

SC Vision Elements	*SC Vision Sub-Elements*	*Written Description (an Example)*	*Desired Rating*
Competitive Advantages	Products/services/ brand	As a differentiated supplier, we simply must bring new and attractive products and services to our customers. We will do this through close customer interactions and active R&D programs	5
	Processes	Our key supply chain processes of 'Strategy', 'Design', Execution' and 'People' are best in class	5

(*Continued*)

Table 3.22 (Continued)

SC Vision Elements	*SC Vision Sub-Elements*	*Written Description (an Example)*	*Desired Rating*
	People	We select, induct, train and assesses our people to ensure high levels of competence. We treat our supply chain people with respect, openness and inclusiveness. We set clear expectations and insist on individual responsibility in achieving them. We make sure that our people have the necessary tools, resources and authorities in order to do their jobs well and achieve their goals	5
	Differentiated customer offer	Our products and services are differentiated on the basis of attractiveness, suitability for purpose, quality, durability and competitive order cycle times. Our customer interfaces include ease of product/service selection, ease of ordering, reliable order status reporting, delivery reliability, and after sales support	5
	Channels to market strength	We desire to operate a coordinated supply chain multi-channel market model providing our customers with a seamless experience. We will consolidate our channels where necessary and carefully select all of our distribution partners	5
	Cost competitiveness	Even though we offer a differentiated customer proposition, we nevertheless desire that our full costs are in the bottom quartile of our industry	5
	Customer loyalty	Retention of existing customers and growth of new ones is our ethos. Strong and working customer relationships are thus essential to our success. We will achieve that by both listening to customers during frequent interactions and faultless delivery of our differentiated offer described earlier	5

Table 3.23 Future Desired State for Each 'People, Regulatory Compliance and Sustainability' Sub-Element (for an Example SC)

SC Vision Elements	*SC Vision Sub-Elements*	*Written Description (an Example)*	*Desired Rating*
People	Numbers of	Every supply chain role in our organisation must have a written and approved position description (PD). The sum of these defines the number of people required to operate our supply chain function. Personnel inflows will be managed based on outflows, relief and training requirements	4
	Skills, capabilities, competencies	These requirements are defined in each PD. People will be selected, inducted, trained and assessed against such criteria	5

SC Vision Elements	*SC Vision Sub-Elements*	*Written Description (an Example)*	*Desired Rating*
	Motivation/energy	Our expectation is that we will have a dedicated cohort of supply chain operatives who are motivated and energetic. This will be achieved through careful selection, induction, training and assessment. Additionally, we treat our supply chain people with respect, openness and inclusiveness. Our supply chain people will have the necessary tools, resources, support and authorities in order to perform at high levels. We desire our reward systems to be mixed between base, short-term and long-term incentives and that it be very competitive	5
	Turnover rates (churn)	Our desire is to have an annual turnover rate of < 5%	1
Regulatory compliance	Occupational health and safety laws	Our standard is that we achieve 100% compliance on all laws and regulations that are relevant to supply chain operations	5
	Competition laws		5
	Environmental laws		5
	Consumer protection laws		5
	Employment laws		5
	Financial reporting rules and regulations		5
Sustainability	Carbon footprint reduction	Our desire is to see year-on-year reductions per unit of output	5
	Energy usage reduction	We seek both energy reductions and migration to cleaner fuels	5
	Waste reduction	To us, waste is evil and we treat it that way	5
	Recycling	Embedded in our new product development processes will be the requirement that our products are fully recyclable	5
	Routing optimisation	We will use advanced analytics, AI and AR in order to optimise all of our transport and distribution activities	5
	Social health	Our desire is to have a growing workforce with a safe, supportive, inclusive and respectful work environment	5
	Economic health	Our overriding objective is survival, and thus our economic health is key to this desire	5

Table 3.24 Overarching SC Goals for Each Vision Element

Vision Element	*Overarching Supply Chain Goals*
Customers	New customers want to buy from us; existing customers are loyal to us
Suppliers	Our suppliers are competitive, stable, ethical and focussed on fulfilling all of our agreed supply chain requirements
Service Providers (SPs)	As per suppliers' goals
Competitors	Our competitors observe all fair trading laws, reporting rules, competition laws and investment policies
Community	Our SC community considers us to be a valued community member
Technical Capabilities	Leading product, process and supply chain technologies protected by IP Advanced design and usage of ICT systems
Competitive Advantages	Our competitive advantages are real and sustained
People	Our supply chain people are our greatest competitive advantage
Regulatory Compliance	100% compliance on all relevant laws and regulations
Sustainability	Our goal is to bring about reductions to our carbon footprint, energy consumption and all forms of waste. We have a key goal of full recyclability of our products. Socially, our goal is employment growth in a safe, respectful, inclusive and supportive work environment. Our economic health goal is for our cashflow and earnings results to be sufficient to ensure our survival. To this end, we will focus on margin management and reductions to all supply chain costs and inventories

strategy details for the first four SC vision elements are shown in Table 3.25. For any real-life SC, such strategy development and description must be completed for the remaining six vision elements as well.

Once the strategies for all ten vision elements have been defined and agreed, it is time to test the strategies for *alignment* with corporate strategies. This can be accomplished by comparing the two side by side as shown for the first four vision elements in Table 3.26. The rating scale used for this example is a simple 5-point Likert scale, where 1 = zero alignment, and 5 = complete alignment. Importantly, SC strategies that are found to align poorly with corporate strategies need to be revisited in order to improve their alignment result. Again, in a real-life SC, the alignment test needs to be carried out for all ten vision elements.

Following the alignment test, it is time to move onto strategy implementation planning. Unfortunately, implementation planning (some people refer to it as 'front end loading') is often poorly done. This results in failed strategy execution and thus non-realisation of SC goals. Implementation planning carried out professionally, thus, is vital to the achievement of defined SC goals. An example implementation plan for ensuring an attractive product portfolio of existing products for the customer vision element is shown in Table 3.27. It is pertinent to point out that each column shown needs to be included in any implementation plan as they are each vital to the success of the plan. The constraint column is pointed out for special mention. Many implementation plans do not include this column, so human resource overloads are not recognised, resulting in missed deadlines.

The implementation plans for more substantial strategies, can be more complex with many implementation steps. An example is shown in Table 3.28. This plan is for the strategy of bringing attractive new products to market. As can be seen, it is a lot more involved (and expensive) than the existing product portfolio clean up shown in Table 3.27.

The last major part of SC strategy setting is that of conducting a thorough risk analysis including a critical risk management plan.

Table 3.25 Overarching SC Strategies and Strategy Details for the First Four Vision Elements

	Overarching Supply Chain Strategies	*Supply Chain Strategy Details*
Customers	To complement our existing product range and by utilising an active R&D program, we will bring new and attractive products to market and deliver them faultlessly to our customers. Tired products will be retired. Our customer engagement processes will be active and constructive. Our customer interface processes will ensure a seamless customer experience	Our customer strategy will be supported by high levels of delivery reliability (98% DIFOTEF), competitive order cycle times with a supply chain that is responsive, flexible and nimble to changing customer requirements and at the same time resilient to supply chain shocks. High levels of customer collaboration and a demand driven planning and scheduling approach facilitated by a digital supply chain are also key components of our customer strategy
Suppliers	We will engage and work with suppliers who offer attractive products on competitive order cycle times and deliver them to a 98% DIFOTEF performance level	We require our suppliers to offer JIT supply solutions on a Vendor Owned Stock (VOS) basis. For JIT to work successfully our strategy is to have high levels of supplier collaboration enabled by Cloud computing, analytics, AI and high visibility of all key supply chain information (demand, supply, inventories, capacities, constraints, plant working ratios, lead times, operating rosters, operations plans and schedules). To reduce our time to market for new products, our strategy is to have our suppliers achieve pre-qualification by making them part of the new product development schedule
Service Providers (SPs)	We will engage and work with service providers who offer leading edge solutions and have a solid reputation for delivering them on-time, on-budget and to-promised functionality	Our strategy for service providers is operate a fully digital model where that is feasible. Supply chain service providers include agents, brokers, freight forwarders, transport providers, warehouse providers, ICT providers, software providers, consultants, auditors, recruiters and trainers. Service providers will be engaged either with turn-key contracts or Service Level Agreements (SLAs)
Competitors	We will continuously develop and improve our competitive advantages such that they are superior to our competitors	

Table 3.26 Testing Alignment of Defined SC Strategies Against Corporate Strategies (First Four Vision Elements)

	Corporate Strategies	*Our Overarching Supply Chain Strategies*	*Alignment Score*
Customers	Our strategy is to be fully customer focused. We will collaborate with customers to design products they want and need, and we will then develop those products and services. We will ensure a one-stop shop and offer seamless experiences for our customers. Our customers will learn to know that we cherish them	To complement our existing product range and by utilising an active R&D program, we will bring new and attractive products to market and deliver them faultlessly to our customers. Tired products will be retired. Our customer engagement processes will be active and constructive. Our customer interface processes will ensure a seamless customer experience	5
Suppliers	Our strategy is to segment our buy requirements and to also segment our supplier base. We will then best match our buy requirements with the segmented suppliers	We will engage and work with suppliers who offer attractive products on competitive order cycle times and deliver them to 98% DIFOTEF	3
Service Providers (SPs)	An increasing component of our non-core activities will be performed by our service providers. It will be important to have active engagement and collaborative processes in place to ensure this works successfully	We will engage and work with service providers who offer leading edge solutions and have a solid reputation for delivering them on-time, on-budget and to-promised functionality	2
Competitors	Legal competitive intelligence gathering is a necessary strategy to ensure we stay ahead of our competition	We will continuously develop and improve our competitive advantages such that they are superior to our competitors	4

Table 3.27 Example Implementation Plan for 'Attractive Product Portfolio' Strategy of the Example SC's Customer Vision Element

Vision Element	*Vision Element Strategy*	#	*What*	*Who*	*How Long (Days)*	*How Much ($,000)*	*When Start*	*Sequence*	*Constraint (Who)*
Customers	Attractive Product Portfolio	1	In consultation with customers, agree on a definition of just what an 'attractive product' is. Record the result	SCL, SML, SCA1, CustRep (25 off)	14	23.6	1-Aug	1	
		2	Review current product portfolio and nominate products for removal that are either no longer attractive and/or unprofitable	SCL, SML, SCA1	14	21.8	20-Aug	2	
		3	Seek approval to remove nominated products from product mix offered	SML	7	4.3	9-Sep	3	SML
		4	Remove approved products	SML	2	1.2	19-Sep	4	SML
		5	Plan next steps	SCL, SML, SCA1	7	10.9	21-Sep	5	
		6	Prepare and present monthly progress report	SCL, SML	2	2.5	1-Oct	6	
				Direct Labour Cost Total =		**64.3**			

SCL = Supply Chain Leader, SML = Sales/Marketing Leader, SCA1 = Supply Chain Analyst #1, CustRep = Customer Representative

Table 3.28 Implementation Plan for 'Bringing New Products to Market' Strategy of the Example SC's Customer Vision Element

Vision Element	*Vision Element Strategy*	*What*	*Who*	*How Long (Days)*	*How Much ($,000)*	*When Start*	*Sequence*	*Constraint (who)*
Customers	Bringing Attractive New Products to Market	Using earlier definition of an 'attractive product', develop proposals for 5 new products within the current product mix breadth	SCL	10	6.15	20-Aug	1	R&DL, SCL, SML, SCA1
			R&DL	30	24.81			
			R&DA	30	10.38			
			SML	10	6.15			
			SCA1	15	4.90			
		Using the same definition, develop proposals for 2 new products outside current mix breadth	SCL	4	2.46	19-Sep	2	SML
			R&DL	12	9.92			
			R&DA	12	4.15			
			SML	4	2.46			
			SCA1	6	1.96			
		Test proposals with customer groups	R&DL	2	1.65	1-Oct	3	SCL, SML
			R&DA	2	0.69			
			SML	2	1.23			
			CustRep (25 off)	0.25	1.83			
		Prepare business cases for each proposal	SCL	7	4.31	3-Oct	4	
			SML	7	4.31			
			R&DL	14	11.58			
			R&DA	21	7.27			
			SCA1	14	4.58			

Seek approval for the 7 proposals	R&DL	2	1.65	24-Oct	5	
	SCL	1	0.61			
	SML	1	0.61			
Prepare a detailed 'project brief' for each approved proposal, including nature, extent and timing of supplier involvement. Project briefs must include design for service, design for quality and design for sustainability requirements	R&DA	21	7.27	26-Oct	6	R&DA
	SCA1	7	2.29			
	SupRep (5 off)	1	1.39			
Assign and induct dedicated R&D teams to each project brief and provide clear signal to commence the work	R&DL	3.5	2.89	16-Nov	7	
	SCL	3.5	2.15			
	SML	3.5	2.15			
Using professional project readiness and assurance methodology, carry out new product R&D on each approved project	R&D Teams	260	4,725	19-Nov	8	
Prepare monthly progress report	R&DA	2	8.31	19-Dec	9	
Present monthly progress report	R&DL	0.5	4.96	21-Dec	10	
	Direct Labour Cost Total =		**4,870**			

SCL = Supply Chain Leader, SML = Sales/Marketing Leader, SCA1 = Supply Chain Analyst #1, R&DL = R&D Leader, R&DA = R&D Analyst, CustRep = Customer Representative, SupRep = Supplier Representative

Table 3.29 Risk Analysis Rating Scale

Risk Criteria Scale						
		Rating				
Risk Criteria	**Likelihood (Probability)**	**1** Rare Frequency greater than 5 years	**2** Unlikely Frequency every 2 to 5 years	**3** Possible Frequency every 1 to 2 years	**4** Likely Frequency 2 to 3 times per year	**5** Almost certain Frequency ≤ monthly
	Consequence (Impact)	**1** Minor injury or environmental incident Business cost < $25k <1 hr delay No damage to equipment	**2** Significant injury Minor environmental incident Business cost $26k to $250k 1 to 10 hr delay Minor equipment damage	**3** Serious injury Serious environmental incident Business cost $0.25M to $2.5M Moderate equipment damage 10 to 100 hr Delay	**4** Single fatality Major environmental incident Business cost $2.5M to $25M 100 to 1000 hr delay Major damage to equipment	**5** Multiple fatalities Extreme environmental event Business cost > $25M > 1000 hr delay Future threatened
	Detectability (Failure Features)	**1** Very easy to measure & detect root cause	**2** Easy to measure & detect root cause	**3** Difficult to detect root cause – shut down required to determine	**4** Hidden, unpredictable and root cause not obvious	**5** Improbable that root cause can be determined

Table 3.30 Risk Category Matrix

Risk Category Matrix							
			Impact				
			Low	*Minor*	*Moderate*	*Major*	*Critical*
			1	*2*	*3*	*4*	*5*
Probability	Almost Certain	5	High	High	*Extreme*	*Extreme*	*Extreme*
	Likely	4	Moderate	High	High	*Extreme*	*Extreme*
	Possible	3	Low	Moderate	High	*Extreme*	*Extreme*
	Unlikely	2	Low	Low	Moderate	High	*Extreme*
	Very Rare	1	Low	Low	Moderate	High	High

Table 3.31 'High' and 'Extreme' Risk Summary for an Example SC

Risk #	*Risk*	*Probability*	*Impact*	*Risk Score*	*Risk Rating*
1	Competitive rivalry intensifies leading to customer demand volatility and market share loss	3	4	7	Extreme
2	Customer relationships deteriorate leading to loss of collaborative efforts	1	4	5	High
3	Unreliable suppliers lead to SKU stock-outs	2	4	6	High
4	Major shock disrupts the supply chain (e.g. earthquake, tsunami, cyclone, pandemic, conflict)	2	5	7	Extreme
5	Digital supply chain project loses the project lead	2	4	6	High
6	High turnover of supply chain personnel due to dissatisfaction	1	4	5	High
7	Competitor poaches our supply chain leader	2	4	6	High
8	Funding for 85% of supply chain improvement projects is withdrawn	2	4	6	High

The first step in conducting a risk analysis is to identify all possible risks to the SC and then to rate them. The rating process requires a scale, and such a risk rating scale is shown in Table 3.29.

Once the risks are identified and rated, it is then possible to categorise them using a risk category matrix as shown in Table 3.30.

Identified risks can be summarised in tabular form as shown in Table 3.31. Importantly, it is crucial that for all 'extreme' and 'high' category risks that a formal risk mitigation plan and contingent action plan be developed and enacted.

For example, for the #1 risk in Table 3.31, the mitigation actions (before the event) might be to focus on building further differentiation features into the SC market offer and to further grow customer relationships via joint collaborations that provide additional customer benefits. Contingent actions (after the event has been triggered) may include improvements to customer service offers, improvements to delivered product and service quality, enhanced service levels agreements and, as a last resort, price competitiveness.

3.4 Testing the efficacy of a developed supply chain *strategy*

Given that the SC overall strategy has been set, how does senior management know that the chosen strategy has indeed the ability to produce the desired result?

To answer this question, the efficacy of the chosen strategy must be tested. There are many ways any given strategy can be tested, ranging from qualitative studies to detailed quantitative studies. Campbell et al. (2000) suggest a sequential testing approach, which, when interpreted for strategy appraisal, can be summarised as:

- i Define the theoretical framework – What is the logic behind this particular strategy? Why would it work? How would it work? What is the logic flow?
- ii Define the testing mechanism – Simulation modelling, qualitative assessment or controlled experiments. Is the testing to be carried out on random individual SC partners (including customers), for example, or on random clusters of partners? Are controlled before and after studies to be conducted?
- iii Conduct initial 'exploratory' studies to ensure the chosen testing technique actually works. Use the results to enhance the testing design
- iv Conduct the study properly, analyse the results, publish and disseminate
- v Adjust the original strategy as necessary in accordance with the test results

If definite time constraints exist, it may be necessary to conduct a qualitative study using an experienced panel of independent SC participants. The questions that such a panel would need to answer for each strategy under consideration are:

- i Is the strategy aligned with the overall corporate strategy and consistent with its neighbouring SC partners' strategies?
- ii Does the chosen strategy take into account the current and expected future business or industry environment? Does it respond to that environment in a desirable way?
- iii Given the logic of the strategy, what is its forecast outcome? Is this forecast acceptable? What complications might arise?
- iv Can the strategy be adequately resourced give the SC's known resource availability? All types of resources need to be considered here (i.e. people, skills, competencies, research capabilities, plant, process, products, equipment, services, money and time)
- v Is the risk profile of this strategy acceptable?
- vi How long will it take to fully implement this strategy? Is that an acceptable time frame?
- vii Is it a *proven* solution – Will it actually work? Has it worked elsewhere? Are there critical success factors for this strategy that are known?
- viii Will SC personnel buy into this strategy? Will it be supported or opposed? If opposed, is a major undertaking required to overcome such opposition?

Such questions need to be answered truthfully and supported with evidence wherever possible. If the answers to the questions are negative (undesirable), then the strategy needs to be revisited and modified such that the cause(s) of negative responses to the questions are removed. If that cannot be done, then serious consideration should be given to replacing the strategy.

3.5 Supply chain *strategy* pitfalls and obstacles

The obstacles to successful strategy implementation are many and varied but can be grouped into *technical* factors and *cultural* (or social) factors.

Technically based obstacles include:

i The appropriateness of the chosen strategy. Is it doable, will it actually deliver the desired result? If the strategy is weak, then it will be an obstacle to itself
ii The level (detail and comprehensiveness) of pre-launch planning carried out. If insufficient, then the risk of cost and time overruns are high. It may even fail because of this, and many implementations indeed have failed because of poor planning
iii Level of competent resourcing. If this is deficient, then implementation obstacles can be expected
iv Degree and quality of training in the 'new way' of the strategy. Again, if deficient, then implementation problems will be experienced
v Use of a reliable implementation and follow-up process such as the 'Project Readiness and Assurance' process. If non-existent, then the implementation cannot be adequately managed
vi The communications process used. If communications are infrequent and/or unreliable, then numerous obstacles will be encountered ranging from people not knowing what is going on, or not knowing what to do next, to sinister types filling the communications gap with aspersions

Culturally based obstacles include:

i The perceived *sense* of the strategy. If this is low, then SC people may well be indifferent to it. It may even come to be considered optional
ii Personnel involvement in the development of and planning for strategy implementation. High levels assure ownership of the strategy, low levels the opposite
iii Personnel attitude and levels of commitment. If high, then chances of implementation success are also high. If low, then the opposite
iv Leadership quality and presence. Competent and present (available, 'on-the-ground') leaders will assure success. Incompetent and absent (e.g. always travelling) leaders assure chaos
v Silo mentality. Unfortunately, many functionally aligned organisations suffer from this mentality – internal organisational functions (or silos) struggle to cooperate or collaborate with one another. This can lead to a number of dysfunctional behaviours which in turn can impact negatively implementations efforts
vi The existence of 'toxic' behaviours. From time to time, regrettably, SC leaders have to deal with individuals who are:

  a *Uncooperative* (unhelpful, disobliging, unaccommodating, unamenable, unreasonable, unwilling, recalcitrant)
  b *Argumentative* (disagreeing, quarrelsome, bickering, wrangling, captious, contrary, cantankerous, dissentient)
  c *Antagonistic* (aggressive, belligerent, pugnacious, combative, contentious, confrontational)
  d *Hostile* (dead set against, opposed, averse, ill-disposed, unsympathetic)
  e Using *work avoidance* tactics (soldiering/malingering)
  f *Malevolent* (malicious, spiteful, evil-minded, bitter, evil-intentioned, poisonous, venomous, malignant, rancorous, vicious, vindictive, vengeful)
  g *Stonewalling* (deliberate blocking or delaying of ideas and initiatives)

h Exhibiting *malicious compliance* (agreeing to an idea or initiative publicly whilst attacking and denigrating it privately. A classic example of this is when people in a meeting commit freely to do something and then walk out of the room and do absolutely nothing about it)
i Demonstrating *fleeting commitment* (everything is optional, people may state that observance of standards and delivery of goals is important in front of their boss; however, such commitment rapidly fades if the boss is distant or there is no consequence management and/or peer pressure influences them negatively)
j *White anting* (subtle sabotage, the act of undermining in order to bring down from within)
k *Acting corruptly* (acting dishonestly, deliberately breaking organisational rules and/or engaging in illegal or unethical behaviour)

For strategy implementation to be completed successfully, therefore, any of the previously mentioned *technical* and *cultural* issues that exist within any given SC must be dealt with proactively and assertively before and, if necessary, during implementation. People exhibiting toxic behaviours, for example, simply need to be removed from the organisation.

3.6 Chapter 3 case study – supply network strategy testing

The case study for this chapter concerns the use of mathematical programming to help select a preferred supply network strategy for bioethanol and sugar production in South Africa (Mutenure et al., 2018).

Bioethanol is considered an important sustainable fuel and marketable product for three main reasons: (i) it is an atmospheric carbon neutral fuel because whilst CO_2 is released from its combustion, that CO_2 release is offset because as the biomass grows, it absorbs CO_2; (ii) it has shown growth in demand year-on-year with favourable pricing (the caveat here, of course, is that the COVID-19 pandemic starting in early 2020 dented both bioethanol demand and pricing has suffered for the duration of the virus's impact) and (iii) it has a high cultivation yield and low energy consumption during processing.

In 2017, the researchers for this case were set the specific aim of determining a supply network strategy that maximised earnings before tax whilst trying to satisfy bioethanol and sugar demand using available and allowable biomass resources. It is important to note that in South Africa food crops are not allowed to be used as an ethanol production feedstock.

The SC in question extended through five layers, including sugarcane farming, milling, crushing, cogeneration and fermentation, and lastly, distribution.

Two main strategies were considered:

i Sugarcane producing areas were to remain static. That is, no new farms were to established or existing farms extended
ii Additional sugarcane producing areas were to be added thus increasing sugarcane production capacity. This strategy required capital investment for new growing areas, new processing plant and warehousing capacity and relevant infrastructure expansion

The solving model used by the researchers comprised 48,035 variables (1,220 of them binary) with 11,876 constraints. A 'GAMS CPLEX' solver was used to execute the model.

The results of the model execution clearly identified that for strategy (i), insufficient allowable biomass capacity existed at the time of the study to satisfy all bioethanol demand. Further, for the strategy (i) scenario that returned the best NPV result, whereby *no ethanol at all* was produced, then the estimated Net Present Value (NPV) returned was predicted to be some 7781 × 10^6 ZAR (South African Rand: 1 ZAR equals approximately US$0.06).

However, under strategy (ii), with increased sugarcane production and increased processing and handling capacities, the result changed substantially. For the strategy (ii) scenario that returned the best NPV result, whereby *no limit was placed on ethanol production*, the NPV predicted was some 175,186 × 10^6 ZAR, an increase of 22.5 times that of the best scenario for strategy (i).

It can be seen, therefore, that such strategy testing can result in enlightening outcomes and provide crucial information and knowledge to strategy formulators in their efforts to devise strategies that generate maximum sustainable stakeholder value.

3.7 Review questions

i What three tools can be used to help better understand any given SC's situation?
ii How should such tools be used in order to ensure maximum benefit from their output?
iii What are the basic elements of a sound SC strategy?
iv What are the ten SC *vision* elements?
v What does strategy *alignment* mean? How can it be tested?
vi What are the key elements of a professional *strategy implementation plan*? Can any of them be discarded or ignored? If yes, why so?
vii In conducting a risk analysis, how should risks identified as 'extreme' or 'high' be responded to?
viii Discuss the different techniques and approaches that can be used to test the efficacy of chosen SC strategies

3.8 Assignment topics

(Each to be 3,000 words minimum or about three A4 pages single spaced. Tables and figures are not to be included in the word count.)

i Describe a SC that in your considered opinion has developed and utilises an *excellent suite* of vision, purpose, SC objectives and goals, specific strategies to be pursued in order to achieve said goals, an implementation plan and a full risk and response analysis. What sort of business results does your example SC typically deliver as measured by economic, social and environmental outcomes?
ii From your own research, describe at least three recent major SC risks that have been missed or mishandled by real-life operating SCs. Why did they miss or mishandle the risk? What corrective actions did they take in response to the risk actually materialising?
iii From your own research, describe an example of a *poor strategy selection* made by a real-life SC. What was the result of this poor strategy selection? How did the SC respond to that result?

iv Describe typical SC *strategy implementation obstacles*. How can SC professionals competently respond if confronted with such obstacles? Use actual examples to justify your answer

3.9 Chapter 3 references

Campbell, M., Fitzpatrick, R., Haines, A., Kinmonth, A., Sandercock, P., Spiegelhalter, D., Tyrer, P. (2000). Framework for Design and Evaluation of Complex Interventions to Improve Health. *BMJ*, 321, pp. 694–696.

Mutenure, M., Čuček, L., Egieya, J., Isafiade, A., Kravanja, Z. (2018). Optimisation of Bioethanol and Sugar Supply Networks: A South African Case Study. *Clean Technology and Environmental Policy*, 20, pp. 925–948.

4 Developing an elegant supply chain *design*

4.1 What you will learn in this chapter

The chapter starts with examples of companies competing in SCs that do SC *design* very well.

Next, an overview of the nine SC *design* sub-processes and the metrics that can be used to assess them is presented.

Following that is an in-depth identification of all of the SC *design* sub-processes, down to and including level six.

Consideration of how to go about assessing the quality and completeness of any developed SC *design* is then presented and discussed. Included is an illustration of how highly interconnected the SC *design* sub-processes are. Given the complexity apparent in that illustration, a recommended hybrid *quantitative* and *qualitative* assessment approach is described that can be used when execution schedule time constraints exist.

The chapter case study presents the results of a SC *design* appraisal conducted by researchers from the University of South Australia and the University of Adelaide. One large wine producer operating within the Australian wine industry was considered and assessed on its level of performance in meeting the three sustainability goals of economic, environmental and social performance, using different facility location configurations. Using a mixed integer programming approach, the researchers uncovered some very telling results.

The chapter concludes with a set of review and assignment questions.

4.2 *Design* sub-processes and metrics

For many organisations and SCs, *design* is not something that is treated in a formal or comprehensive way. There may be many examples of disparate design activities in such organisations and SCs within individual functions or projects; however, design as an overarching function or formal process does not usually exist.

On the other hand, there are some excellent examples of organisations and SCs that treat SC *design* with the seriousness and with the front-of-mind attention that it deserves (and in most cases demands). Examples of such organisations include Unilever, Amazon, Inditex, 3M, Apple, Hanes, Walmart and Toyota. Such companies have spent considerable effort and funding over many years perfecting their SC *design* and *execution*. As a result, they are each successful, in concert with their SC partners, in delivering long-run stakeholder value.

In this section, nine SC *design* sub-processes will be introduced, that is *design for*:

i Safety
ii Service
iii Quality
iv Sustainability
v Integration
vi People
vii ICT systems
viii Costs
ix Risks

The level three sub-processes for *design for SC safety* along with metrics used to assess sub-process performance are shown in Table 4.1. As safety is a key licence to operate and social responsibility factor, design quality of the level three sub-processes shown in Table 4.1 is vital to SC longevity.

Similarly, the level three sub-processes and metrics for *design for SC service* are shown in Table 4.2. As can be seen, the level three sub-processes here are crucial for the level of performance actually achieved on SC customer order fulfilment, cost management and sustainability. So again, it is vital to invest the time and effort in getting the *design* for these sub-processes as near perfect as possible.

The next group of level two SC *design* processes is that of *design for quality, sustainability, integration and people*. Each such level two design process and the corresponding level three sub-processes with metrics are shown in Table 4.3. Again, each of the sub-processes shown here are fundamental to levels of SC performance achieved in practice.

The last group of level two SC design processes is that of *SC ICT design, design for SC costs and design for SC risks*. Each such level two design process and the corresponding level three sub-processes with metrics are shown in Table 4.4.

In concluding this introductory section, it must be emphasised that the identified level three SC *design* processes do not exist in isolation to one another; that is, they are highly interrelated, so changes made to one sub-process will have ripple effects on other level three sub-processes. They need to be tested as a system, therefore, towards the end of any SC *design* undertaking. This aspect is considered in detail in Section 4.4.

Table 4.1 Level Three Sub-Processes and Metrics for *Design for SC Safety*

Level Two	*Level Three*	*Metrics Used for Level Three*
Design for safety (as captured in SC Safety Management System)	Design of SC safety policy (see also ISO 45001:2018 OH&S management system)	Approved and ISO safety standard compliant safety policy exists and is up to date
	Design of SC safety planning process to be used	Safety planning & governance procedures exist & are followed
	SC safety implementation process design	Implementation planning and execution procedures exist & are followed
	SC safety measurement, evaluation and improvement process design	Approved set of safety metrics exists, including the process for regular monitoring, reviewing, reporting and improvement
	Design of management review of safety process	Approved management review process exists, and scheduled reviews exist in the SC's management system

Table 4.2 Level Three Sub-Processes and Metrics for *Design for SC Service*

Level Two	*Level Three*	*Metrics Used for Level Three*
Design for service	SC methodology to be used (e.g. push, pull, CPFR, ECR, VMI, VOS, Agile)	Clear methodology choice made + service level, inventory level, lead time and responsiveness tested
	SC physical design	Approved SC physical design exists + service level, inventory level, lead time, responsiveness and cost tested
	Facility layout designs (intra-each-facility)	Intra-facility design exists + service level, inventory level, lead time and responsiveness tested
	Routing optimisation design	Optimised routing design exists + service level, inventory level, lead time and responsiveness tested
	Planning and scheduling methods to be adopted	Planning & scheduling methods chosen + planning cycle time, plan veracity and planning costs tested
	Optimal inventory design	Optimal inventory design exists + days of inventory and frequency of stock-outs tested

Table 4.3 Level Three Sub-Processes and Metrics for *Design for SC Quality, Sustainability, Integration and People*

Level Two	*Level Three*	*Metrics Used for Level Three*
Design for quality	Product/service design & performance	Fitness for purpose, reliability, durability, ease of use, defect free, compliance to specification
	Process design & performance	Cost of quality, reject rate, defect rate
SC design for sustainability	Design for economic performance	Return on capital margin, free cash flow, total supply chain costs
	Design for social performance	Stakeholder and community rating, tenure of relationships, jobs and safety
	Design for environmental performance	Carbon footprint, transport efficiency, licence compliance
SC design for integration	SC strategic integration design	Level of SC partner integration, commonality of SC vision, purpose, goals and strategies
	SC personnel integration design	Cross-enterprise positions, training & personnel sharing
	SC process integration design	Level of integration across strategy, design, execution & people processes
SC people design	SC leadership design	Leaders' performance expectations to meet PD requirements
	SC managerial system design	Right number of suitably qualified, trained & motivated personnel with approved PDs working in an aligned structure
	Desired SC cultural profile design	Attitude, motivation & behavioural standards of staff

Table 4.4 Level Three Sub-Processes and Metrics for *SC ICT design, Design for SC Costs and Design for SC Risks*

Level Two	*Level Three*	*Metrics Used for Level Three*
SC ICT design (ICT plan)	ICT vision, goals and objectives	Index of ICT plan's ability to deliver overarching SC ICT strategy
	SC systems infrastructure design	Index of designs ability to deliver SC ICT infrastructure performance requirements
	Software applications design	Proven solutions, relevant to SC goals, modern, adaptable, supported, affordable
	Process integration support design	Accept inputs from & send outputs to other processes/systems
	Security design	Levels of protection over IP, commercial information, system manipulation & crashes
	Technical support design	Support availability, competence & professionalism
Design for SC costs	SC cost targets design	Fixed and variable cost targets in top decile of best practice
	Design of SC features & practices that will deliver cost targets	Cycle times, waste levels, days of inventory, productivity levels, process integration
Design for SC risks	Design for identification, evaluation and management of all risks	Completeness of risk identification, evaluation thoroughness, response quality

4.3 The *design* sub-processes in more detail

Section 4.2 provided an introduction to SC *design* and presented some key metrics that can be used to assess level three sub-process performance. But what is it that enables such performance? How can SC professionals make it happen? To answer such questions, it is necessary to delve deeper into levels 4, 5 and 6 of the SC *design* process.

To begin, we will consider the first *design for SC safety* sub-process which is that of the *design of SC safety policy* to be used. This level three sub-process has eleven level four sub-processes. The first of these is the safe work methods sub-process, and its elements are shown in Table 4.5. As can be seen, the development of a competent *SC safe work methods design* is no small undertaking, and this is only the first *design of SC safety policy* level four sub-process.

Table 4.5 Levels Five and Six Sub-Processes for *SC Safe Work Methods Design*

Level Four	*Level Five*	*Level Six*
SC safe work methods design process	SC safety standards process	Hazards
		Working at heights
		Personal protective equipment (PPE)
		Security and access
		Safety training
		Fire protection and response
		Equipment isolation, lockout and tagging
		Load tie-down and restraint
		Mobile equipment
		Equipment and machine guarding
		Transportation vehicles and equipment

Level Four	*Level Five*	*Level Six*
	SC job specific hazards, risks and controls process	SC partners SC leader SC planners SC schedulers SC process equipment operators SC mobile equipment operators SC maintenance personnel SC transport drivers (trucks, buses, locos, watercraft, aircraft) SC distribution personnel
	SC job specific safe work instructions	Same level six processes as 'SC job specific hazards, risks and controls process'
	SC job specific standard operating procedures	Same level six processes as 'SC job specific hazards, risks and controls process'

Table 4.6 Levels Five and Six Sub-Processes for *SC Safety Risk Management Process Design, SC Safety Employee Engagement and Communication Design and SC Safety Responsibilities and Accountabilities Design*

Level Four	*Level Five*	*Level Six*
SC safety risk management process design	Identification of all hazards	Work environment Materials, substances Plant and process Manual handling
	Assess and rate risks associated with all hazards	High risk – > potential death, permanent disability Medium risk – > Medical treatment, temporary disability Low risk – > injury requiring first aid treatment
	Control risks	Eliminate Substitute Isolate Design change Procedural instructions Personal Protective Equipment (PPE)
SC safety employee engagement and communications design	Hazards and risks	Engagement and communications process
	Safe work procedures and checklists	*- ditto -*
	Safety standards and audits	*- ditto -*
	Personal safe behaviour expectations	*- ditto -*
	Site induction requirements	*- ditto -*
	Sign-in, sign-out requirements	*- ditto -*
	Emergency procedures	*- ditto -*
SC safety responsibilities and accountabilities design	SC leaders	Demonstrable, active leadership
	SC operatives	Live the safety culture
	SC health and safety manager	Support, example, coach and mentor
	Compliance with work, health and safety statutory regulations	Scheduled audits and corrective actions

The next group of level four sub-processes for the *design of SC safety policy* are those of *SC safety risk management process design, SC safety employee engagement and communication design and SC safety responsibilities and accountabilities design*. The level five and six sub-processes for each are shown in Table 4.6.

Table 4.7 Levels Five and Six Sub-Processes for *Design for the Safe Operation of SC Plant and Equipment* and *SC Hazardous Materials Safe Handling Design*

Level Four	*Level Five*	*Level Six*
Design for the safe operation of SC plant and equipment	Operations	Standard operating procedures (SOPs) Operating manuals Auditing schedules Checklists and corrective action forms
	Maintenance	Inspection schedule Equipment condition standards Inspection results register Corrective action records Mobile equipment procedures Pre-start checklists
	Electrical	Isolation and tagging procedures Temporary equipment installation procedures Electrical repairs and replacements procedures Inspection schedule Equipment condition standards Inspection results register Corrective action records
SC hazardous materials safe handling design	Hazardous materials register Chemical data sheets for all hazardous chemicals Safe handling procedures and warnings PPE requirements Incident and accident reporting requirements	

The next two *design of SC safety policy* level four sub-processes are *Design for the safe operation of SC plant and equipment* and *SC hazardous materials safe handling design*. The level five and six sub-processes for each are shown in Table 4.7.

The next three *design of SC safety policy* level four sub-processes are *employee safety training design, accident and injury prevention design* and *SC safety data, information and dissemination design*. The level five and six sub-processes for each are shown in Table 4.8.

The last two *design of SC safety policy* level four sub-processes are *emergency preparedness design* and *design for safety records and documents management*. The level five and six sub-processes for each are shown in Table 4.9.

The second *design for SC safety* level three sub-process is that of *design of the SC safety planning process to be used*. The level four sub-processes for this are shown in Table 4.10.

The last three *design for SC safety* sub-processes are the level three sub-process of *SC safety implementation process design, SC safety measurement, evaluation and improvement process design and design of management review of SC safety process*. The level four, five and six sub-processes for these are shown in Table 4.11.

To reinforce the point and as it can be seen from Tables 4.5 to 4.11, the proper and comprehensive management of *design for SC safety* is no small undertaking. In practice, it takes strong and persistent leadership, long-term committed resources and the development of a very distinct '*safety culture*' along the entire SC. In the author's direct experience, SC excellence *is underpinned* by SC safety excellence. That is, the discipline of focus,

Table 4.8 Levels Five and Six Sub-Processes for *Employee Safety Training Design, Accident and Injury Prevention Design* and *SC Safety Data, Information and Dissemination Design*

Level Four	*Level Five*	*Level Six*
Employee safety training design	General plant and process induction	
	Specific SC job induction	
	Safety policy	
	Job specific hazards, risks and controls	
	Job specific safety rules	
	Workplace facilities	
	Emergency procedures and expectations	
	Accident, incident and injury reporting requirements	
	Workplace access and security requirements	
	Employee safety training register	
Accident and injury management design	Injury response responsibilities	
	Notification process	
	First aid procedures for all injuries	First aid equipment and location
		Training and when to use
	Medical treatment injuries procedures	
	Rehabilitation procedures	
	Reporting and records requirements	
SC safety data, information and dissemination design	SC safety data collection and storage	
	Data cleansing and validation	
	Data analysis and results reporting	
	SC safety information system (safety portal)	

Table 4.9 Levels Five and Six Sub-Processes for *Emergency Preparedness Design* and *Design for Safety Records and Documents Management*

Level Four	*Level Five*	*Level Six*
Emergency preparedness design	Evacuation procedures	
	Notification of emergency services responders	Service ID, contact position(s), contact method, call numbers, backups
	Medical treatment and assistance procedures	Treatment type, treatment procedures, accreditations required
	Emergency response procedures for differing emergencies	Emergency type, response procedure, notification procedure
	Emergency coordination and communications	Who, what, where, when and how
	Emergency drills	Frequency
		Involvement
		Results and corrective actions
		Record of all drills, results and follow-up actions
	Training and instructions	Schedule and observance
Design for safety records and documents management	Safety policy	
	Safety standards	
	Identification and assessment of hazards and risks	
	Hazard and risk mitigation	
	Safe work instructions and procedures	
	Accidents, incidents, near misses and injuries	
	Safety audits and corrective action results	
	Health monitoring results	
	Training and licensing records	

Table 4.10 Level Four Sub-Processes for *Design of the SC Safety Planning Process to be Used*

Level Three	*Level Four*
Design of SC safety planning process to be used	SC safety vision and purpose SC work health and safety policy SC safety definitions SC safety responsibilities and accountabilities SC safety standards Location specific OH&S requirements Consultation and communication SC safety training SC safety risk assessment Risk register Emergency procedures Site access and security Auditing process and schedule SC safety continuous improvement process SC safety document management

Table 4.11 Level Four, Five and Six Sub-Processes for *SC Safety Implementation Process Design, SC Safety Measurement, Evaluation and Improvement Process Design and Design of Management Review of SC Safety Process*

Level Three	*Level Four*	*Level Five*	*Level Six*
SC safety implementation process design	Review approved safety plan for complexity, timing and resourcing load	Review schedule, list of participants, review and rating process, reporting procedure	
	Identification of key implementation personnel	Identification of those who are specifically responsible, accountable, consulted or informed (using RACI matrix)	Personnel constraint identification
	Develop detailed implementation plan identifying what, who, when, duration, sequence and all constraints	Selection and testing of professional project management techniques to be used	Detailed implementation plan and review schedule
	Assemble implementation teams and give clear signal to start execution	Kick-off meeting schedule and expectation setting	
	Implementation monitoring, reporting and corrective action management	Monitoring and reporting schedule and responsibilities. Corrective action process	
SC safety measurement, evaluation and improvement process design	Selection and testing of *lead* indicators	Safety standards rating process design (comprehensiveness, quality, currency) Auditing process rating design (relevance, quality, inclusiveness, currency)	

Level Three	*Level Four*	*Level Five*	*Level Six*
		Audit schedule on-time completion rate	
		Audit scores design	
		Risk profile rating	
		Management commitment to safety	Demonstrable behaviours
	Selection and testing of *lag* indicators	Lost time injury frequency rate	
		Medical treatment injury frequency rate	
	Safety improvement process design	Process rating (workability, relevance, reliability, results, user appraisal)	
		Design of process usage rating	Frequency of use % coverage of all safety improvement attempts
Design of management review of SC safety process	Process for managing frequency of management safety reviews against standard	Data collection and analysis. Analysis results reporting	
	Process to ensure reviews are scheduled into managers' diaries	Automated link between safety management system and leaders' diaries	
	Process for monitoring and controlling on-time completion rate of management safety reviews	Fixed agenda item for leadership team meetings	
	Quality rating of management safety reviews	Independent and accredited auditor reviews	
	Attendance rate at scheduled safety reviews by individual managers	Data collection and analysis. Analysis results reporting	

attention to detail, persistence, setting workable standards and ensuring compliance to the same, once mastered for safety, can be applied to every other SC excellence factor.

The second level two SC *design* sub-process is that of *design for service* which in turn has six level three sub-processes. As operating SC professionals well know, this SC activity underpins the everyday customer service result experienced in real life. The first level three sub-process in this set is the selection and approval of the SC operating methodology to be used: 'push', 'pull', 'collaborative planning, forecasting and replenishment (CPFR)', 'efficient consumer response (ECR)', 'theory of constraints (TOC)', 'vendor managed inventory (VMI)', 'vendor owned stock (VOS)' or 'Agile'. The levels four, five and six sub-processes for this are shown in Table 4.12.

The next two level three sub-processes for *design for service* are *selecting and setting the SC physical design* and the *intra-facility layouts* within that overall design. The level four and five sub-processes for each of these are shown in Table 4.13.

Table 4.12 Level Four, Five and Six Sub-Processes for *Selection and Approval of the SC Operating Methodology to be Used*

Level Three	*Level Four*	*Level Five*	*Level Six*
Selection and approval of the SC methodology to be used (e.g. push, pull, CPFR, ECR, TOC, VMI, VOS, Agile)	Make to order using lean SC methodology	Fulfil order from SC input materials. Make at a rate that matches order levels and buffer requirements	Actual order levels Required service level Unit capacities Lead times Scheduling rules Calculation of constraint resource buffer size Calculation of despatch buffer size Calculation of timing and level of material launch into SC recognising constraint resource capacity, order levels, order due dates, lead times and buffer level requirements (i.e. MRP)
		Fulfil order using WIP stock. Make at a rate to match actual + forecast order levels	Same as previously mentioned except need order level forecasts also
	Make to stock using push methodology	Make at maximum rate until stopped by material shortages	Scheduling rules to be applied
	Make to stock using lean SC methodology	Set maximum stock levels and make at maximum rate until stock limits reached	Scheduling rules and maximum stock levels to be applied

Table 4.13 Level Four and Five Sub-Processes for *Selecting and Setting the SC Physical Design* and the *Intra-Facility Layouts*

Level Three	*Level Four*	*Level Five*
Selecting and setting SC physical design	Design location of plant, facilities, transport, distribution, ICT infrastructure, administration, suppliers, service providers and customers	Map GPS coordinates per instance Geography (places and people) Terrain (e.g. mountainous, flat, boggy, rocky, fertile) Altitude Climate (e.g. temperature, rainfall, humidity, weather events) Distances Availability and cost of suitable land for plant and facilities Transport and services infrastructure (e.g. roads, rail, canals, airfields, electricity, fresh water, gas, telecommunications)
Facility layout designs (intra-each-facility)	Design of flows and equipment layout	Materials Services Information Personnel Reverse SC flows
	Workstation layout design	Location in chain Detailed layout of workstation Number of operators and station capacity

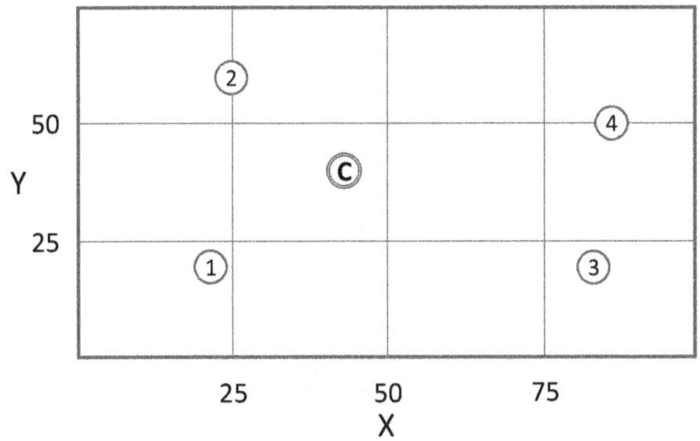

Figure 4.1 Centre-of-Gravity Facility Location Method Example

A method that can be used to assist with SC physical design is the simple *centre-of-gravity* method. Take the case of Crusty Bakery, a baking business that wants to determine the best location for a new bakery facility in a new town. Research has identified the location of the suppliers of the four necessary materials used by Crusty Bakery and the number of truck runs required from each per month:

Supplier 1 is located at map coordinate (20,20) and requires 35 truck runs per month. Supplier 2 is located at map coordinate (25,60) and requires 42 truck runs per month. Supplier 3 is located at map coordinate (80,20) and requires 20 truck runs per month. Supplier 4 is located at map coordinate (85,50) and requires 15 truck runs per month.

The *centre-of-gravity* location for the new Crusty Bakery facility can thus be calculated using the equations:

$$C_X = \Sigma L_{ix} \times R_i/\Sigma R_i \text{ and } C_Y = \Sigma L_{iy} \times R_i/\Sigma R_i$$
$$C_X = 20 \times 35 + 25 \times 42 + 80 \times 20 + 85 \times 15 = 4625/112 = 41.3$$
$$C_Y = 20 \times 35 + 60 \times 42 + 20 \times 20 + 50 \times 15 = 4370/112 = 39$$

The location of the four suppliers and the calculated *centre-of-gravity* location (point 'C') for this example are shown in Figure 4.1.

The fourth and fifth level three sub-processes for *design for service* are *routing optimisation design* and *planning and scheduling methods to be adopted*. The level four and five sub-processes for each of these are shown in Table 4.14.

A routing minimisation technique that can be used for complex networks whereby every node on the network has to be connected or visited is the minimal-spanning tree problem. Figure 4.2 shows a network of seven nodes, each of which are connected by arcs. The numbers shown on each arc are the travel distances in this case. The problem to be solved is how to service each of the nodes whilst traversing the least distance. The process is fairly straightforward with the steps involved being:

1 Select a starting node and connect it to the nearest node (i.e. least distance away) in the network

Table 4.14 Level Four and Five Sub-Processes for *Routing Optimisation Design* and *Planning and Scheduling Methods to be Adopted*

Level Three	*Level Four*	*Level Five*
Routing optimisation design	Specific SC design attributes	Source node locations
		Intermediate locations
		Destination locations
		Route costs/unit between each node
		Route capacities between each node
		Route transit durations between each node
		Supply capacities
		Customer demand levels
	Optimisation criteria	Minimise total routing costs
		Maximise total routing throughput
		Minimise total routing duration
		Special considerations for specific routes
	Optimisation technique	Mathematical programming
		Simulation
		Machine learning
Planning and scheduling methods to be adopted	Long-range SC plans (1 to 25 years)	Organisational business plan
		SC master plan (including high-level MRP)
	Medium-term SC plans (1 to 52 weeks)	Strategic SC optimisation
		SC budget plan
		SC Sales and Operations Plan (S&OP) (including MRP)
	Short-term SC plans (1 hour to 7 days)	SC master schedule (including MRP)
		SC unit schedule

2 Find and connect the next nearest node that is not already connected. If there is more than one nearest node, then select one of the nearest nodes arbitrarily
3 Repeat Step 2 until all nodes are covered

The bold lines shown in Figure 4.2 represent the solution to the minimal-spanning tree problem.

The last level three sub-process for *design for service* is that of *optimal inventory design*. The level four sub-processes for each of these are shown in Table 4.15.

The third level two SC *design* sub-process is that of *design for quality* which has two level three sub-processes: *product/service design and performance* and *process design and performance*. The level four and five sub-processes for each of these are shown in Table 4.16.

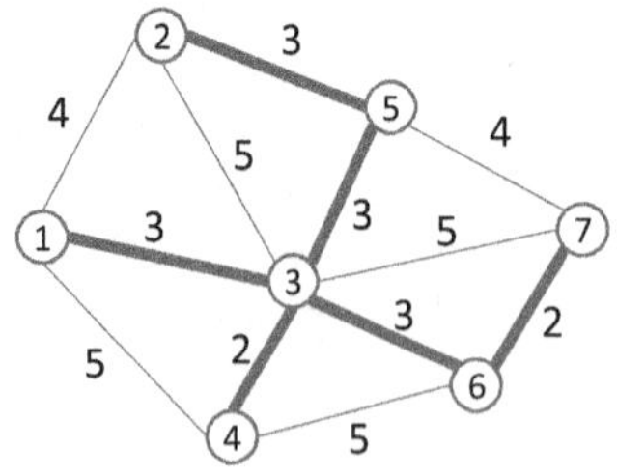

Figure 4.2 Solution to the Minimal-Spanning Tree Problem

Table 4.15 Level Four Sub-Processes for *Optimal Inventory Design*

Level Three	*Level Four*
Optimal inventory design	Set days-of-inventory targets/SKU and overall Set max stock outs tolerable/SKU/time duration Using selected SC operating methodology (Table 4.12), set stocking rules per inventory location Set days-of-inventory targets/SKU/SC location Set calculation methods for determining levels of all strategic inventory buffers (e.g. constraint resource and despatch buffers) On a per customer basis, agree and set ownership of SC stock (e.g. VMI or VOS) Detail the design of the inventory reduction continuous improvement process to be used For units using kanbans, show how to calculate size and number of kanbans Detailed design of inventory visibility system (end-to-end SC)

Table 4.16 Level Four and Five Sub-Processes for *Product/Service Design and Performance* and *Process Design and Performance*

Level Three	*Level Four*	*Level Five*
Product/service design & performance	Product/service planning	Detailed definition of customer needs, wants and desires for product and service quality Product/service required specific characteristics, attributes and targets and tolerances (e.g. standard, simple designs, with reduced component parts, lower production costs, within ± tolerance and less than Six Sigma defective)
	Product/service plan deployment design	Technical design parameters at a component level
Process design & performance	Process planning	Process technology selection Specific equipment selection Facility configuration designs Process capability specification (C_p required to meet product quality requirements and how it is to be achieved) Process capability with shift specification (C_{pk}) Performance target assurance check
	Production planning	Design of the overall production control system to be used (who, what, where, how) Design of production planning systems to be used Implementation plan

The fourth level two SC *design* sub-process is that of *design for sustainability* which has three level three sub-processes: *design for economic performance, design for social performance and design for environmental performance.* This set really is a three-legged stool type, because if any given SC neglects any one of the legs, the stool falls over. The level four and five sub-processes for each of these are shown in Tables 4.17a, 4.17b and 4.17c.

The fifth level two SC *design* sub-process is that of *design for integration* which also has three level three sub-processes: *SC strategic integration design, SC personnel integration design* and *SC process integration design*. The level four and five sub-processes for each of these are shown in Table 4.18.

Table 4.17a Level Four and Five Sub-Processes for *Design for Economic Performance*

Level Three	*Level Four*	*Level Five*
Design for *economic* performance	Attractive new product to market process design	Customer engagement process, new product prototype development & testing, new product certification process, new product launch process
	Attractive portfolio of existing products management process	Existing product assessment process, product retirement process, records currency process
	Competent service design	Design for quality, delivery performance, lead time, responsiveness, flexibility and resilience
	Days of inventory (DOI) levels design	Specific inventory management process design (e.g. JIT, TOC, algorithmic)
	SC total costs design	ABC unit cost design for materials, services, conversion, distribution, overheads
	Price: cost ratio design to give aim margins	Target margins per SKU and material selection, conversion and delivery process design needed to deliver the target margins
	SC asset intensity design	Comparable analysis plus material selection, conversion and delivery process design needed to meet asset intensity targets

Table 4.17b Level Four and Five Sub-Processes for *Design for Social Performance*

Level Three	*Level Four*	*Level Five*
Design for *social* performance	Job availability	Job growth targets and overall business design to achieve same followed by SC designs to match
	Rates of pay design	Process technology designs required to meet productivity targets
	Working hours design	Working hours to be set by SC ordinary hours per week standard
	Roster systems design	Roster system design to be set through extensive consultation and employee roster type preferences that meet required throughput levels
	Working conditions design	Globally consistent working condition designs that comply with set ILO standards
	Occupational health and safety design	Globally consistent OH&S designs that comply with ISO standards
	Disruption standard for non-SC people trying to live their lives	Air pollution, water pollution, soil pollution, noise pollution, visual pollution (e.g. tall buildings, bright lights), traffic congestion, extravagant use of land, deforestation management processes
	Community engagement design	Objectives, who, what, where, when, budget spend, engagement and communications processes
	Community joint initiatives design	Community council set-up design, management of collaborative initiatives process
	Compliance with all social regulations design	Regulations awareness and training processes. Strategies to be adopted to ensure compliance. Auditing processes

Table 4.17c Level Four and Five Sub-Processes for *Design for Environmental Performance*

Level Three	*Level Four*	*Level Five*
Design for *environmental* performance	Minimum environmental impact design	Air quality management process (e.g. water vapour, CO, CO_2, Methane, NOx, SOx, SF6, NF3, fluorocarbons and hydrofluorocarbons) Water quality management process Soil quality management process Noise pollution management process Visual pollution management process
	Traffic load design	Air, land and water traffic density targets, air and noise pollution targets and tactics to be applied to meet the targets
	Land usage footprint design	Units produced/hectare of land used or customers satisfied/hectare of land used
	Transport type and routing design	Air, road, rail, sea, canal, pipeline, hypertube, volumes shipped/mode/route and environment impacts against targets
	Compliance with all environmental regulations design	Regulations awareness and training processes. Strategies to be adopted to ensure compliance plus auditing process to be followed

Table 4.18 Level Four and Five Sub-Processes for *SC Strategic Integration Design, SC Personnel Integration Design* and *SC Process Integration Design*

Level Three	*Level Four*	*Level Five*
SC strategic integration design	The case for SC integration (expected benefits: i.e. why)	Rationale, expected outcomes, costs, obstacles, risks, overall benefits and returns on effort and cost
	SC partner integration design	Which SC partners specifically will integrate (who)
	SC integration methodology design	Integration methodology to be adopted (what, how, when)
	Written, agreed SC integration charter	Objectives and targets, who, what, when, where, how, scope, reporting details
	Review and improvement process design	
SC personnel integration design	The case for SC personnel integration (expected benefits: i.e. why)	Rationale, expected outcomes, costs, obstacles, risks, overall benefits and returns on effort and cost
	SC personnel integration design	Which roles?
	Detailed position descriptions for each design component	Duties, expected outcomes, responsibilities, authorities, adjacent role relationships
	Location and reporting details design	
	Review and improvement process design	
SC process integration design	The case for SC process integration (expected benefits: i.e. why)	Rationale, expected outcomes, costs, obstacles, risks, overall benefits and returns on effort and cost
	SC process integration design	Which processes?
	ICT support design	
	Responsibility and authority design	
	Review and improvement process design	

Table 4.19 Level Four and Five Sub-Processes for *SC Leadership Design, SC Managerial System Design* and *SC Cultural Profile Design*

Level Three	*Level Four*	*Level Five*
SC leadership design	Design of process to be used by leaders to help in setting direction	
	Design of process to be used by leaders to ensure all resources are used effectively	
	Leaders' technical capabilities requirements design	
	Leaders' social capabilities requirements design	
	Leaders' selection/induction process design	
	Leaders' training/development process design	
	Leaders' performance requirements design	
	Leaders' effectiveness review process design	
	Leaders' rewards and recognition system design	
SC managerial system design	Aligned organisational design	
	Administration processes design	
	Team formation process design	
	Personnel cost structure design	
	Digital personnel support systems design	
	Review and improvement process design	
Desired SC cultural profile design	Principles design	Trust
		Fairness
	Values design	
	Behavioural design	
	Motivation and commitment design	
	Review and improvement process design	

The sixth level two SC *design* sub-process is that of *SC people design* which also has three level three sub-processes: *SC leadership design, SC managerial system design* and *SC cultural profile design*. The level four and five sub-processes for each of these are shown in Table 4.19.

The seventh level two SC *design* sub-process is that of *SC ICT design* which has six level three sub-processes. The first three of these are *ICT vision, goals and objectives, SC systems infrastructure design* and *software applications design and selection*. The level four and five sub-processes for each of these are shown in Table 4.20.

The second three of level three *SC ICT design* sub-processes are *process integration and support design, security design* and *technical support design*. The level four and five sub-processes for each of these are shown in Table 4.21.

The last two level two SC *design* sub-processes are *design for SC costs* and *design for SC risks*. The level four, five and six sub-processes for each of these are shown in Table 4.22.

This concludes the identification of all of the key SC *design* sub-processes. They are extensive, and it is important to emphasise that they take considerable time and resources to do properly. However, if done poorly or not at all, the result will lead to sub-optimal SC performance. It is an unfortunate condition that many people in organisations seem to prefer to rush to execution with design pushed into the background (Choi et al., 2001). This most often results in mismatches, bottlenecks, stopgap measures and compromises. Worse, it is usually very difficult to correct such unsatisfactory conditions once a SC has been set in place. As Varsei and Polyakovskiy (2017) conclude in their study described in this chapter's case study: 'It appears the unpurposeful design could have negative environmental, social and economic implications'. The importance of doing SC *design* well

Table 4.20 Level Four and Five Sub-Processes for *ICT Vision, Goals and Objectives, SC Systems Infrastructure Design* and *Software Applications Design and Selection*

Level Three	*Level Four*	*Level Five*
ICT vision, goals and objectives	Vision statement	E.g. our vision is to have a fully digital, customer centric SC
	Goals and targets	Functionality, availability, response times, change flexibility, ease of access, visibility, security, access, failure recovery, backup, scalability, cost
	Infrastructure policy and performance requirements	On-premises, cloud, hybrid, service type
	SC ICT strategies	Level of differentiation, performance, cost
	ICT technology adoption plan	Progression timeline, projects and milestones
	ICT disaster response and recovery plan	Detailed disaster management plan, including identification of possible disasters, expected frequency and typology
	ICT security policy and plan	
SC systems infrastructure design	Facilities design	Networks, hubs, buildings, wireless
	Hardware design	Computers, servers, desktop and mobile appliances
	Software design	Operating systems, applications and portals
	Network design	Connectivity design, speed, carrying capacity, reliability and cost
	User interface design	Desktop, tablet, smartphone, screen designs
	Cloud usage design	
Software applications design and selection	Operating system design	Cost/performance analysis, supply contract details, costs
	SC applications design	Functionality, integration, reliability, change flexibility, user acceptance, value for money
	SC software on Cloud design	
	SC applications development method design	E.g. in-house or standard commercial or tailored commercial or hybrid?

Table 4.21 Level Four and Five Sub-Processes for *Process Integration and Support Design, Security Design* and *Technical Support Design*

Level Three	*Level Four*	*Level Five*
Process integration support design	SC process integration design	Who, what, when, where and how
	Integration hardware requirements design	Stand-alone, shared hardware, Cloud partnership model
	Integration software requirements design	Shared SC software/applications
	SC integration collaboration process design	Collaboration requirements and how they are to be met
Security design	SC IP security design	
	SC data security design	
	SC data access protocol design	
	SC data backup design	
	SC data and systems recovery design	
Technical support design	Technical support availability design	Availability target and testing
	Technical support ease of access design	Time taken to access target and testing
	Quality of support service	Competence Language skills Attitude Professionalism
	Technical support in-house/outsource design	
	Technical support cost design	

Table 4.22 Level Four, Five and Six Sub-Processes for *Design for SC Costs* and *Design for SC Risks*

Level Three	*Level Four*	*Level Five*	*Level Six*
Design for SC costs	SC costs targets	Input costs design	Materials, services, utilities
		Conversion costs design	Cost/unit/SKU, cost/operating hour/SKU
		Distribution costs design	Transport cost design Service centre cost design Warehouse cost design
		ICT systems costs design	Infrastructure, Cloud cost, software, ICT personnel, training and upgrade costs
		SC personnel costs design	Total cost/classification, cost/hr/classification, cost/unit of output
	SC features & practices that will deliver cost targets	Days of inventory (DOI) design	DOI target and assurance process
		SC cycle times design	Cycle time design per process per SKU
		SC waste elimination design	Waste target & design to achieve
		SC productivity design	Productivity target and design to achieve
		Overall equipment effectiveness (OEE) design	OEE target and design to achieve
		SC process integration design	Why, who, how, extent
Design for SC risks	Identify, evaluate and manage all risks	Risk identification process design	
		Risk rating process design	
		Risk ranking process design	
		Risk preventative action process design	
		Risk contingent action process design	

and *before* execution, cannot be stressed enough because it is so crucial to long-term SC performance.

The next section considers how to go about conducting an assessment of the quality and completeness of a developed SC *design*.

4.4 Assessing the quality and completeness of a developed supply chain *design*

Working for many months, even years, to complete a SC *design* is one thing, but will the completed design deliver the goods? That is, will it, if implemented, provide the level of supply chain excellence sought by the SC partners? This is a vital question to answer comprehensively and objectively.

Numerous studies have been conducted on SC network design, and many models have been developed to test aspects of a given SC *design*. Such approaches typically apply analytical methods or simulation or a combination of the two (Chiadamrong and Piyathanavong, 2017), (Fattahi et al., 2020), (Garg et al., 2004), (Varsei and Polyakovskiy, 2017).

Also, a number of specific quantitative methods have been developed to help with SC *design*. Examples include mathematical programming (especially goal programming) on network models for route and facility location optimisation, queuing theory and transportation and transhipment models (Melo et al., 2009).

Whilst such models used as testing regimes are commendable and encouraged, they do require time and effort to develop, test and validate. Additionally, the number of linkages involved in modelling all of the SC *design* processes together, as shown in Figure 4.3, would require the development of a very complex analytic and/or simulation model. This would involve a very long development time followed by additional time in conducting the research necessary to confirm the linkage directionality and to establish the strength of each linkage. Plus, the model would have to be maintained on an ongoing basis to ensure its currency. In short, the time required to do all this may not be available at a time when SC leaders are wanting to move onto execution.

A hybrid approach that uses both *qualitative* and *quantitative* methods, resulting in a more time efficient method for SC process efficacy testing, is therefore recommended.

The *quantitative* part involves the use of pre-existing quantitative methods such as the use of mathematical programming (especially goal programming) on network models for route, transport type and facility location optimisation, queuing theory for shopfront designs and transportation and transhipment models for distribution considerations on those parts of the SC *design* that can be developed within an acceptable time frame.

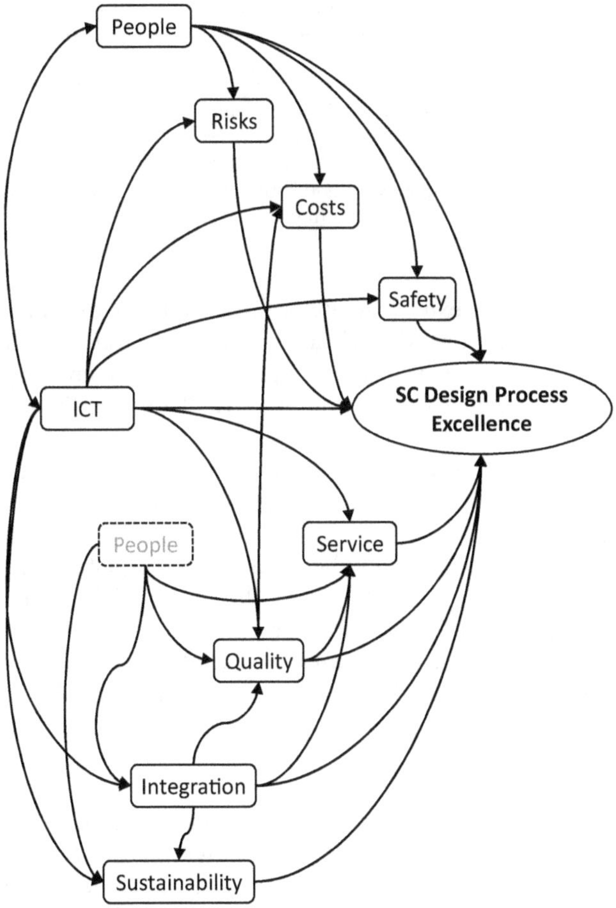

Figure 4.3 Interconnectivity of SC Design Processes

Table 4.23 Rating of the *Design for Safety* Level Three Sub-Processes

Level Two Process	#	*Level Three Processes*	*Level Three Metrics*	*Rating*
Design for Safety	1	Design of safety policy	Approved and ISO safety standard compliant safety policy exists and is up to date	5
	2	Design of SC safety planning process	Safety planning & governance procedures exist	4
	3	SC safety implementation process design	Implementation planning and execution procedures exist	3
	4	SC safety measurement, evaluation and improvement process design	Approved set of safety metrics exists, including the process for regular monitoring, reviewing, reporting and improvement	5
	5	Design of management review of safety process	Approved management review process exists, and scheduled review requirements exist in the SC's management system	2

Table 4.24 Rating Scale for Assessing SC Sub-Processes

Rating	*Characteristics*
1	Very low quality (not relevant, poor design, not fit for purpose)
2	Low quality (barely relevant, basic design, fit for purpose shortfalls)
3	Average quality (relevant, functional design, marginally fit for purpose)
4	High quality (very relevant, robust design, fit for purpose)
5	Very high quality (highly relevant, advanced design, fit for purpose plus)

This doesn't preclude to any extent the use of further quantitative methods as time permits. As the *quantitative* part provides valuable, even critical, input to the *qualitative* part, it needs to be completed *before* the *qualitative* appraisal. It is recommended therefore that the *quantitative* model building work be done in parallel to the SC *design* development work.

The *qualitative* part involves the rating of each level four, five or six SC sub-processes (the design ones in this case) against their assessed ability to perform to the desired level for the level three metrics.

An example of how to do this, for the example of *design for safety*, is illustrated in Table 4.23. The rating scale used for this approach is shown in Table 4.24.

It is important to point out that the following questions must be answered truthfully and objectively when determining the rating scores. Also, the scoring should be carried out by an experienced and unbiased cross-functional team and not by an individual or by a group whose members have responsibility for the process under review. The questions to be asked and answered are:

i What is the *quality* of the process features (all levels) that have been included and described to achieve the desired level of the target metric? (Table 4.24 rating scale can be used to assign a 1 to 5 rating score). *Quality* here refers to the *relevance* of the design features, the design *strength* (poor to excellent) and the *fitness for purpose* of the proposed process design

ii What actual evidence exists to support this rating? (Research studies, plans, process maps, process descriptions, proven interrelationships, outputs from the *quantitative* part, any verification tests that have been conducted)
iii What is the comparative for this sub-process design? That is, how does it stack up against what competitors or SCs in like industries are doing and the results they are achieving?

Such an appraisal needs to be carried out for each of the nine SC *design* sub-processes. Any that are rated < 4 should be revisited to determine if their rating needs to be higher and, if yes, what is to be included in the process design to ensure that is achieved.

Lastly, the reliability of this assessment process needs to be tested after the SC *design* has been implemented: Do the ratings given at the time of the assessment correlate with the results actually achieved from the operating SC? If not, then the causes of such discrepancies need to be identified and careful modifications made to the process. Low correlations mean that either the ratings were inappropriate or the developed design was insufficient in delivering the desired outcome or both. Either way, SC leaders need to know the cause of any such low correlations and then to initiate corrective actions.

4.5 Chapter 4 case study – developing a sustainable supply chain *design*

The topic of this case study is an investigation conducted by Varsei and Polyakovskiy (2017) into the question of SC sustainability improvements to the SC design of an Australian wine producer named ABC.

The Australian wine industry is of considerable economic and social benefit to the people of Australia. In economic terms and from its annual 1.2 billion litres of wine production, the wine industry in Australia contributes an economic value to the country that is in excess of AUD 45 billion per year. In social terms, the industry in total provides jobs for more than 160,000 people.

The researchers set out to determine if the SC *design* in use by ABC at the time of the 2014 study could be improved in order to:

i Reduce costs, both fixed and variable
ii Improve environmental performance, specifically reduce CO_2 emissions
iii Improve social outcomes, specifically employment levels and regional impact

ABC is a top-10 Australian wine producer, and information was collected on the company via a series of interviews and data collection activities during 2014. Details were collected about the in-use SC *design*, production and transportation costs, volumes of production and customer demand by region. Equipment capital costs were obtained from other suppliers, whilst transport details were obtained from logistic providers. CO_2 emission factors for different transport types were calculated using the 'International Wine Carbon Calculator Protocol Version 1.2'. Social coefficients were calculated based on the impact on employment and GDP in each potential bottling plant location.

At the time of the study, ABC had one large winery plant and bottled the wine into 750 ml glass bottles at an adjacent bottling plant. From the bottling plant, the wines were shipped to the major demand centres in Queensland, New South Wales, Victoria,

Table 4.25 Optimisation Results for the Three Objective Functions

Scenario #	*Configuration Bottling Plant Location(s)*	*Total Costs/Year AUD × 10^6*	*CO_2 Emissions tonnes/year*	*Social Coefficients*
1	In use (as-is, adjacent to wine plant)	44.6	4936	0.67
2	Large bottling plant in Melbourne	37.5	2215	0.78
3	Large bottling plant in Melbourne, small one in Sydney	38.7	1780	0.85
4	Large bottling plant in Adelaide, small one in Brisbane	47.3	7151	0.97

Source: Varsei and Polyakovskiy (2017)

Tasmania, South Australia, Northern Territory and Western Australia. Transportation was mostly by road (trucks) except for Victoria where it was shipped by rail.

The researchers concluded that due to its proximity to grape suppliers, moving the winery to another location was not feasible. The bottling plant, however, could feasibly be moved to other locations because the wine could be transported satisfactorily in bulk between the wine plant and in-market bottling facilities. This included rail and ship transport for the bulk wine shipments.

Using CPLEX 12.6, an augmented constraint multi-objective, mixed-integer mathematical program was used to solve the optimisation problem for the three objective functions of *costs*, *CO_2 emission* and *social impact.*

The results are very telling. For the first four scenarios considered, the researchers clearly illustrated that the in-use SC *design* (scenario one) was obviously not developed to optimise any of the three objectives. This conclusion can be appreciated from observation of Table 4.25.

As can be seen, the in-use configuration (scenario one) is the second worst for cost and emissions and worst for social result. Much better options are scenarios two and three.

It depends on the priority that the owners of the ABC wine producer place on cost performance of course; however, scenario three seems to offer the most favourable and balanced outcome configuration of the four considered in this group.

This case study illustrates well the power and usefulness of *quantitative* methods in assessing SC *design*. This statement can only grow in relevance as the requirement for SCs to meet global sustainability goals intensifies.

4.6 Review questions

- i What are the nine sub-processes that are important to include when developing a relevant and competitive SC *design*?
- ii What metrics can be used to assess the quality of a SC *design for service*?
- iii Given that the SC *design* sub-process are interrelated, what does that mean for SC professionals in developing *SC designs*?
- iv What is it about design for safety attributes which, if mastered, can be used to enhance the other SC *design* sub-processes?
- v What typical outcomes are experienced when SC personnel rush to execution before a robust SC *design* is completed?
- vi What techniques can be employed to test the efficacy of a developed SC *design*? What interdependencies need to be considered in the usage of such techniques?

vii Using this chapter's case study as an example, comment on the power and usefulness of quantitative methods in the assessment of SC *designs*

4.7 Assignment topics

(Each to be 3,000 words minimum or about three A4 pages single spaced. Tables and figures are not to be included in the word count.)

i From your own research, list at least three companies that make use of some of the nine SC *design* sub-processes. Describe how the companies involved use such sub-processes in practice, and identify the results they have achieved from doing so

ii From your own research, identify a SC that reliably achieves high levels of SC service performance (i.e. very competitive DIFOTEF, order cycle times, responsiveness and flexibility). Describe in detail, using diagrams to aid your description, the design features of this SC that underpins such reliable service performance

iii Describe a SC where the process of SC *design* was done very poorly or not at all. What were (or are) the consequences of this? (Consider economic, environmental and social consequences separately) Were the consequences addressed? If yes, were shortfalls fixed? If yes, how so and what was the result?

iv From your research, describe a situation whereby a researcher or SC professional has used either quantitative or qualitative (or both) methods to assess the efficacy of a SC *design* (or at least a part of a SC *design*). How did they go about it? What results did they achieve? (*Hint: many research papers have been written on SC network design.*)

4.8 Chapter 4 references

Chiadamrong, N., Piyathanavong, V. (2017). Optimal Design of Supply Chain Network Under Uncertainty Environment Using Hybrid Analytical and Simulation Modelling Approach. *Journal of Industrial Engineering International*, 13, pp. 465–478.

Choi, T. Y., Dooley, K. J., Rungtusanatham, M. (2001). Supply Networks and Complex Adaptive Systems: Control Versus Emergence. *Journal of Operations Management*, 19(3), pp. 351–366.

Fattahi, M., Govindan, K., Maihami, R. (2020). Stochastic Optimization of Disruption-Driven Supply Chain Network Design with a New Resilience Metric. *International Journal of Production Economics*, 230, pp. 1–16.

Garg, D., Narahari, Y., Viswanadham, N. (2004). Design of Six Sigma Supply Chains. *IEEE Transactions on Automation Science and Engineering*, 1(1), pp. 38–57.

Melo, M. T., Nickel, S., Saldanha-da-Gama, F. (2009). Facility Location and Supply Chain Management – A Review. *European Journal of Operational Research*, 196(2), pp. 401–412.

Varsei, M., Polyakovskiy, S. (2017). Sustainable Supply Chain Network Design: A Case of the Wine Industry in Australia. *Omega*, 66, pp. 236–247.

5 Achieving supply chain *execution* excellence

5.1 What you will learn in this chapter

The chapter begins with an introduction to the seven level two SC *execution* sub-processes, including a listing of all of the level three sub-processes and their key metrics.

Following this is a description of the level four, five and six sub-processes for all of the SC *execution* sub-processes.

Two detailed examples are also included.

The first provides a detailed description of a *sales and operations planning* process, including demand, capacity, make level, despatches, management of safety stock, facility loading and utilisation and days of inventory calculation.

The second illustrates the main activities that need to be undertaken to competently execute a *project readiness and assurance* process.

An end of chapter case study is included which considers the 2020 top three and the 2020 'Masters' group as appraised by Gartner's 'Supply Chain Top 25 for 2020' rating process. Reasons for each candidate's rating are presented and discussed.

Finally, a series of review questions and written assignments are included.

5.2 Supply chain *execution* sub-processes and metrics

SC *execution* is the action part of SCM; that is, execution is mostly about the physical '*act of doing*'. SC *strategy* and *design* are also about doing; however, most of the work in those two processes is cerebral in nature. Some people refer to SC *strategy* and *design* as more about 'conceptualisation', whereas SC *execution* is more about 'actualisation'. SC *execution*, thus, is the SC process the delivers actual results to customers every day. The performance of the *execution* process very much shapes customer appraisal of and attitudes towards the SC.

SC *execution* excellence requires expertise and competence in the management of all of the level three, four, five and six SC *execution* sub-processes that are presented and discussed in this chapter.

To begin the coverage, each of the seven level two SC *execution* sub-processes will be identified along with the corresponding level three sub-processes and the level three metrics that can be used to monitor performance and set improvement targets.

The seven level two SC *execution* sub-processes are:

i New products and services delivery to customers
ii Customer service

iii Order fulfilment
iv Managing returns
v Enabling suppliers
vi Collaboration, cooperation and integration
vii SC execution support systems

The first three level two SC *execution* sub-processes along with their level three sub-processes and metrics are shown in Table 5.1.

The fourth and fifth level two SC *execution* sub-processes along with their level three sub-processes and metrics are shown in Table 5.2.

The last two level two SC *execution* sub-processes along with their level three sub-processes and metrics are shown in Table 5.3.

As for the SC *strategy* and SC *design* sub-processes and in concluding this introductory section, it must be emphasised that the identified level three SC *execution* processes are highly interrelated. For any given SC, the strength of these relationships needs to be established and taken into consideration during their operationalisation.

Table 5.1 First Three Level Two SC *Execution* Sub-Processes with Metrics

Level Two	*Level Three*	*Metrics Used for Level Three*
New products, services to market	New product/service performance specification	Index of alignment between performance specification and customer desires, performance specification feasibility index
	Engagement of suppliers and service providers	Formal engagement process exists, record of meetings and attendance, record of agreements reached
	Authentication of suppliers and service providers	Authentication process exists, cycle time duration, number of errors reported
	New product/service planning process execution	Planning cycle time, number of errors from new product/service planning process execution
Customer service	Publish offer	Approved offer exists, is published and available to all customers, offer-review score
	Customer contracts, agreements	Agreed and approved contracts/agreements exist, customers' score
	Capture orders	Index of ease of ordering and order entry errors
	Order Management (OM)	Duration of OM cycle, OM error rate
	During and after sales service	Timeliness, quality, relevance
Order fulfilment	SC planning	Planning cycle duration, forecast accuracy, index of optimisation, planning cost
	Source	Sourcing lead time, stock-outs, days of raw materials inventory, quality and cost of sourced products/services
	Make	Days of WIP and finished good inventory, quality of made products/services, constraint unit(s) utilisation rates, make costs
	Deliver	% DIFOTEF, order cycle times, transport times, wait times, delivery costs

Table 5.2 Fourth and Fifth Level Two SC *Execution* Sub-Processes with Metrics

Level Two	*Level Three*	*Metrics Used for Level Three*
Managing returns	Returns request notification and initial response	Response times to notifications, notifications error rate
	Return routing	Time taken to decide, time taken to advise, error rate
	Receive return	Time taken from notification to receival and receival to customer confirmation of receival
	Return treatment decision	Time taken to decide, time taken to advise, error rate
	Return processing	Processing rate, processing cycle time, level of waste
	Continuous improvement	Approved process exists, improvement rate achieved per Key Performance Indicator (KPI) addressed
	Return costs	Total cost of handling returns + cost of each return activity/unit
Enabling suppliers	Categorise supply base	Formal categorisation exists, index of usefulness
	Make/buy/outsourcing decisions	Time taken to decide, index of decision quality, actual results achieved
	Supplier appraisal	Approved appraisal process exists, index of usefulness, cost of appraisal
	Rationalisation of the supply base	Time taken to decide, index of decision quality
	Developing supplier potential	Approved process exists, improvement rate achieved per KPI addressed
	Early supplier involvement in new product development	Approved process exists, effectiveness index
	Supplier negotiations	Approved win-win process exists, improvement rate achieved (e.g. quality, costs, inventory levels), profit sharing index
	Supplier relationships (including partnerships)	Approved relationship plan exists, extent of plan implementation, index of performance improvement achieved
	Managing supplier performance	% DIFOTEF, delivery cycle times, days of inventory, responsiveness, costs

Table 5.3 Sixth and Seventh Level Two SC *Execution* Sub-Processes with Metrics

Level Two	*Level Three*	*Metrics Used for Level Three*
Collaboration, cooperation, integration	SC relationship management	Index of relationship strength, index of results delivered from improved relationships
	SC information sharing	Information sharing index (including information 'width and depth')
	SC performance measurement	Overall SC % DIFOTEF, delivery cycle times, days of inventory, responsiveness, costs
	SC people effectiveness execution	Index of effectiveness, effectiveness appraisal schedule, actual results achieved by teams and communities of practice

Level Two	*Level Three*	*Metrics Used for Level Three*
SC execution support systems	ICT plan execution	Number of strategic ICT projects launched, on-time delivery performance of ICT projects, on-budget performance of ICT projects, delivery of promised functionality performance of ICT projects, index of execution progress against individual ICT plans
	ICT infrastructure execution	Infrastructure gaps, infrastructure performance index
	Software/applications execution	Customer satisfaction level with customer interface systems, suitability index for each SC control system, in-service quality, cost
	e-Commerce execution	% of SC partners using, rating of experience, cost, benefits
	Process integration execution	Index of integration extent, index of integration effectiveness
	ICT project management	Constraint-based process exists, % on-time delivery, % to functionality promise, performance to $ budget

5.3 The supply chain *execution* sub-processes in more detail

The first level two SC *execution* sub-process is that of new products and services delivery to market. This level two sub-process has four level three sub-processes as shown in Table 5.4. The four level three sub-processes shown here are set-up processes to enable deployment to order fulfilment (i.e. the actual doing) which is shown as the last level five sub-process in the table.

All of the sub-process outlined in Table 5.4 need to be configured and applied in any given SC for the metrics shown in Table 5.1 to be achieved.

The second SC *execution* sub-process is that of customer service. This level two sub-process has five level three sub-processes as shown in Table 5.5. Importantly, the level five features identified need to be included and operationalised for any SC seeking performance leadership. There is a lot of work, of course, in ensuring all of the level four and five requirements are competently completed and patiently bedded down. SC professionals, however, enable SC process excellence via the completion of such work to a high standard.

The third SC *execution* sub-process is that of order fulfilment. This level two sub-process has four level three sub-processes the first of which is shown in Table 5.6.

At this point, it is appropriate to consider a more detailed example of a completed SC sub-process. The example chosen is that of Sales and Operations Planning (S&OP). S&OP is a very critical SC process as it sets the entire execution 'game plan' for the SC.

The example consists of a portion of a SC containing three partners. Two aggregated product groups are included with the identities Alpha and Beta. The two product groups

Table 5.4 Level Three, Four, Five and Six Sub-Process of New Products and Services Delivery to Market Sub-Process

Level Three	*Level Four*	*Level Five*	*Level Six*
New product/ service performance specification	Input from new product/ service design		
	Confirmation that installed process capability and capacity can successfully deliver to the performance specification	Calculation of C_p and C_{pk}	Data collection, cleansing & analysis + results response process
		Calculation of Operating Equipment Effectiveness (OEE)	Data collection, cleansing & analysis + results response process
		Calculation of working ratio and realistic capacity per hour, per day, per week, per month	SC flow design, constraint identification, queue times and processing times per unit
Engagement of suppliers and service providers	Confirmation that suppliers/service providers have capabilities to meet the performance specification	Assessment of all necessary certifications, accreditations and performance records	Data collection, data cleansing and data verification
	Confirmation that suppliers/service providers have the capacity to meet supply rate and timing requirements	Checking of suppliers working ratio and capacity results	Request for cleansed and verified data from suppliers
	Preparation and publication of agreed and approved service level agreements	Service level agreements drawn up and ratified by signature of authorised persons	Service level agreements negotiated and agreed
Authentication of suppliers and service providers	Confirmation of supplier/service provider identity, address details and key contact personnel	Request for cleansed and verified identity data from suppliers	
	Provision of necessary sign-on and access permissions	Nominations approved and entered into security register	Sign-on and access permission policy developed and approved
New product/ service planning process execution	Input from new product/ service planning design		
	Execution of the designed planning process	Sales & Operations Plan (S&OP)	Demand plan, source plan (MRP), make plan, despatch plan, inventory plan
		Distribution plan	Transport plan Warehousing plan
		Unit schedules	Scheduling rules, unit capacities, constraints, lead times, due dates
		Plan appraisal and approval	
		Plan deployment to order fulfilment	

Table 5.5 Level Three, Four and Five Sub-Processes of Customer Service Sub-Process

Level Three	*Level Four*	*Level Five*
Publish offer	Build attractive website with ease of access	Customer engagement, input from specialists
	Products and services offered clearly described	Build, test and syndicate system
	Value propositions clearly presented and explained	Prepare presentation and deliver face-to-face
	Information is up to date and complete	Enter into auditing and review schedule
	Provision of help features	Customer engagement, input from specialists
	Customer queries feature	- *ditto* -
	Regular review and upgrades	Enter into auditing and review schedule
Customer contracts, agreements	Relationship building focus	Develop actioning plan from *strategy* and *design* inputs, engage iteratively with customers
	Collaboration emphasis	Agree on collaboration initiatives with customers, set and agree on the schedule
	Tailored for specific customers	Individual customer engagement and inputs
	Flexible to changing customer circumstances	Translate flexibility strategy and design into action
	Periodically reviewed and modified	Enter into auditing and review schedule
Capture orders	Encouragements to place order	Collect data on special offers, discounts, personalised offers
	On-line customer chat feature	Customer engagement, input from specialists
	Ease of ordering	- *ditto* -
	Order entry error checking	- *ditto* -
	Order confirmation and link to order status reporting	- *ditto* -
	Tax invoice issue (if applicable)	Trigger invoice launch
Order Management (OM)	Capture and storage of all order data	Confirmation of availability and error checking
	Order history function	Confirm ICT system
	Order details check against product catalogue	Run order validity test
	Order fulfilment inventory and capacity check	Confirm whether order will be filled from finished goods, work in process or raw material stock
	Setting of order delivery target and link to order status reporting	Calculate lead time based on stock type and processing path
	Review of order management performance	Enter into auditing and review schedule
During and after sales service	Before order placed	Ease of systems access Ease of human contact Competence and reliability of service staff Honouring all follow-up commitments
	After order placed	Same as before order placed plus reliable order status reporting and in-service advice

Table 5.6 SC Planning Processes of the Level Two Order Fulfilment Sub-Process

Level Three	*Level Four*	*Level Five*
SC Planning	Sales and Operations Plan (S&OP) (1 month to 36 month forward plan)	Confirm product/service aggregations Determine actual order levels Determine forecast order levels Confirm capacity forecasts Confirm product/service lead times Confirm product/service costings (ABC costing) Determine optimal SC routings Calculate optimal loading profiles for operating units Calculate opening and closing inventory positions Confirm feed availability and supply rates Calculate feed requirements Determine despatch forecasts Link plan result to financial forecast
	Master Production Schedule (MPS) (1 day to 30 day product/service plan)	Input is S&OP What specific products/services are to be produced? How many specific products/services are to be produced? When are those specific products/services to be produced? Output to Materials Requirements Plan (MRP)
	Materials Requirements Plan (MRP)	Input is MPS Confirm product/service demand levels Confirm product/service routings and lead times Confirm product inventory positions Using specific products' Bill of Materials (BOM) and previously mentioned information, calculate feed requirements per time period
	Unit Schedule	Input is MPS and MRP Confirm unit scheduling rules Confirm unit capacities Confirm unit set-up times For each production unit and compliant with scheduling rules, assign product make sequence, amount and aim production time For each make sequence, determine constraint resource(s) Calculate feed launch amounts and timings Calculate size of constraint and despatch buffer (safety stocks) Non-constraint units inventory managed by kanbans

are processed through three facilities, a1, b1 and a2, each of which are owned by one of the three SC partners. The layout can be visualised as shown in Figure 5.1.

The example S&OP explained here has been deliberately kept simple to aid in explanation. However, it needs to be noted that the methodology described would be the same for a SC containing, for example, ten partners, 150 product groups and 15 processing facilities.

A S&OP is really a high-level calculated forecast for a range of key SC parameters such as demand levels, feed material orders, feed material deliveries, feed inventory levels, processing (or 'make') capacity, make levels, work-in-process (WIP) and finished goods inventory levels and finished goods despatches.

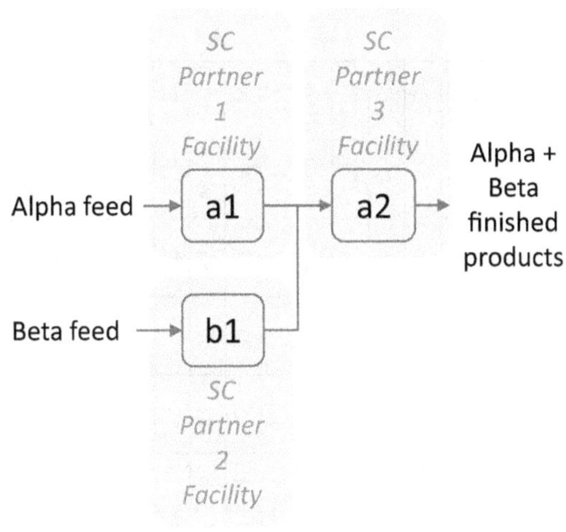

Figure 5.1 Product Flow Path for Sales and Operations Plan Example

Such parameters for the chosen example are shown in Table 5.7.

The time horizon and the time 'buckets' (time intervals) used in practice will vary from SC to SC; however, typically the time horizon used is between three to 18 months and the time buckets usually days or weeks. For this example, a three-month time horizon has been chosen along with weekly time buckets. For simplicity, the lead time for feed orders is one week.

Observation of Table 5.7 identifies a number of features:

i There is quite a level of variability evident in the key parameters of demand and make capacity. This makes stability along the chain difficult to achieve as evidenced by the variations that can be seen in feed order levels, feed inventory levels, make levels and finished goods inventory levels

ii For the Alpha product group, backlogs (late deliveries) evident early in the plan are removed as sufficient feed material and make capacity exist to allow that to happen, even given the level of variability that is evident. From week #5 on, all demand commitments are met

iii For the Beta product group, the situation is quite different. Nine of the 13 weeks are forecast to be in backlog. This is caused by a shortage of capacity as the make units for product Beta run at maximum capacity from week #2 to the end of the plan. This is an unacceptable situation, so the SC planner responsible would need to identify which specific facility or facilities are in overload and propose solutions

iv The high level of interrelationships between the parameters in the plan can be appreciated by observing weeks 6, 7 and 8 for the Alpha product. Week 6 finishes with a high level of closing finished goods inventory, so to reduce this back down to aim levels, the make level for week 7 is reduced to 78 units. This reduction in make level causes an excursion in week 8's opening feed inventory level (less make uses less feed) which is not corrected until week 11

Table 5.7 13-Week Sales and Operations Plan for Product Groups Alpha and Beta

Product Group	*Month – >*	*January*					*February*				*March*			
	Week # – >	*1*	*2*	*3*	*4*	*5*	*6*	*7*	*8*	*9*	*10*	*11*	*12*	*13*
Alpha	Demand (units)	120	114	125	108	112	99	102	123	115	120	127	112	104
	Backlog (late deliveries)	25	17	11	9	0	0	0	0	0	0	0	0	0
	Feed inventory safety stock aim	7	7	7	6	6	5	5	6	6	6	6	6	5
	Opening feed inventory	10	7	7	7	0	6	0	49	50	17	6	6	0
	Feed delivery (units)	120	120	127	116	136	104	126	129	79	118	123	112	132
	Feed order placed (units)	120	127	116	136	104	126	129	79	118	123	112	132	123
	Opening feed inventory position	-15	-4	-2	6.1	24	11	24	55	14	15	2.3	5.9	28
	Make capacity (units)	123	120	127	125	130	127	121	128	123	129	123	125	127
	Make level (units)	123	120	127	123	130	110	78	128	111	129	123	118	92
	Closing feed inventory	7	6.6	6.8	0	5.6	0	49	50	17	6	6.3	0	40
	Opening finished goods inventory	5	0	0	0	6.1	24	35	10	15	12	21	17	22
	Finished goods despatches	128	120	127	117	112	99	102	123	115	120	127	112	104
	Finished goods safety stock aim	14.5	13	14	12	11	9.9	10	12	12	12	13	11	10
	Closing finished goods inventory	0	0	0	6.1	24	35	10	15	12	21	17	22	10
Beta	Demand (units)	84	85	82	80	84	81	84	85	81	82	87	81	87
	Backlog (late deliveries)	0	0	0	1	0	2	3	8	11	2	2	7	2
	Feed inventory safety stock aim	4	4	4	4	4	4	4	5	5	4	4	4	4
	Opening feed inventory	5	10	4	4	0	4	4	4	5	5	4	4	4
	Feed delivery (units)	89	78	80	78	85	80	79	82	90	82	82	86	87
	Feed order placed (units)	78	80	78	85	80	79	82	90	82	82	86	87	85
	Opening feed inventory position	10	3.3	2.1	1	1.2	1.2	-4	-6	2.6	2.2	-3	2.4	2.5
	Make capacity (units)	90	84	80	82	81	80	79	82	90	82	82	86	87
	Make level (units)	84	84	80	82	81	80	79	82	90	82	82	86	87
	Closing feed inventory	10	4	4	0	4	4	4	5	5	4	4	4	4
	Opening finished goods inventory	2	2	1	0	1	0	0	0	0	0	0	0	0
	Finished goods despatches	84	85	81	81	82	80	79	82	90	82	82	86	87
	Finished goods safety stock aim	8	9	8	8	8	8	9	9	9	8	9	9	9
	Closing finished goods inventory	2	1	0	1	0	0	0	0	0	0	0	0	0

The logic behind such a plan is quite complex, and there are many calculations involved. For instance, the 'make level' logic for this example can be described using a pseudo-algorithm as:

```
   If demand + backlog ≥ make capacity, then
   If opening feed inventory + feed delivery ≥ make
capacity, then
   Make level = make capacity
   Else
   Make level = opening feed inventory + feed delivery
   Else
   If opening feed inventory + feed delivery < make
capacity, then
   Make level = opening feed inventory + feed delivery
   Else
   If closing finished goods inventory < finished goods
safety stock aim, then
   Make level = make capacity
   Else
   Make level = demand + backlog - closing finished goods
inventory + finished goods safety stock aim
   End
   End.
```

The second part of this S&OP example concerns the question of operating facilities and their loading levels. Table 5.8 illustrates this. As can be seen, the facilities a1 and a2 are able to fulfil their loading requirements with infrequent weekly overloads that can be caught up with subsequent week underloads. Facility b1, however, tells a different story. Here the facility is heavily loaded in all weeks and indeed overloaded in weeks 9 and 13. The SC planner in this case would recommend the working of overtime on the overloaded weeks plus the addition of additional capacity, either through facility throughput continuous improvement efforts or, if that is not possible, via the installation of new equipment.

The last part of the example is that of inventory results from the plan. This can be seen by observing Table 5.9. The results show a very favourable days of inventory (DOI) result of 0.19 days for the 13-week plan. The high utilisation rate of the operating facilities is one reason the SC inventories are so low as priority was placed in fulfilling demand rather than building safety stocks. It could be argued that such a low level of inventory is indicative of a very lean SC and thus one that is intolerant of any SC shocks. The use, therefore, of reliable SC resilience building methods that include a comprehensive SC shocks risk analysis in order to calculate necessary safety stocks and to arrange emergency alternate feed supply options cannot be overemphasised. The SC resilience builder approach, as explained in Section 2.6 of this book, is therefore recommended.

The last three order fulfilment sub-processes are those of '*source*', '*make*' and '*deliver*'. The level four sub-processes that need to be executed here are shown in Table 5.10.

The fourth level two SC *execution* sub-process is that of *managing returns*. This sub-process has seven sub-processes as shown in Table 5.11. For their particular SC, SC professionals need to establish each level four and level five sub-process shown and instigate

Table 5.8 13-Week Facility Capacity and Loading Plan

Facility	*Month – >*	*January*					*February*				*March*			
	Week # – >	*1*	*2*	*3*	*4*	*5*	*6*	*7*	*8*	*9*	*10*	*11*	*12*	*13*
a1	Hours available	80	80	80	80	80	80	80	80	80	80	80	80	80
	Hours/unit alpha	0.65	0.65	0.65	0.65	0.65	0.65	0.65	0.65	0.65	0.65	0.65	0.65	0.65
	Units of alpha required	123	120	127	123	130	110	77.6	128	111	129	123	118	92
	Hours used on alpha	79.95	78	82.6	80	84.5	71.2	50.4	83.2	72.3	83.9	80	76.7	59.8
a2	Hours available	120	120	120	120	120	120	120	120	120	120	120	120	120
	Hours/unit alpha	0.5	0.5	0.5	0.5	0.5	0.5	0.5	0.5	0.5	0.5	0.5	0.5	0.5
	Units of alpha required	123	120	127	123	130	110	77.6	128	111	129	123	118	92
	Hours used on alpha	61.5	60	63.5	61.6	65	54.8	38.8	64	55.7	64.5	61.5	59	46
	Hours/unit beta	0.7	0.7	0.7	0.7	0.7	0.7	0.7	0.7	0.7	0.7	0.7	0.7	0.7
	Units of beta required	84	84	80	82	81	80	79	82	90	82	82	86	87
	Hours used on beta	58.8	58.8	56	57.4	56.7	56	55.3	57.4	63	57.4	57.4	60.2	60.9
b1	Hours available	40	40	40	40	40	40	40	40	40	40	40	40	40
	Hours/unit beta	0.47	0.47	0.47	0.47	0.47	0.47	0.47	0.47	0.47	0.47	0.47	0.47	0.47
	Units of beta required	84	84	80	82	81	80	79	82	90	82	82	86	87
	Hours used on beta	39.5	39.5	37.6	38.5	38.1	37.6	37.1	38.5	42.3	38.5	38.5	40.4	40.9

Table 5.9 13-Week Inventory Plan

Product Group	*Facility*	*Month – >*	*January*					*February*				*March*				*Daily Average*
		Week # – >	*1*	*2*	*3*	*4*	*5*	*6*	*7*	*8*	*9*	*10*	*11*	*12*	*13*	
Alpha	a1	Closing feed inventory	7	7	7	0	6	0	49	50	17	6	6	0	40	15
	a2	WIP inventory	6	3	8	6	0	7	2	0	0	0	0	0	3	3
		Finished goods inventory	0	0	0	6	24	35	10	15	12	21	17	22	10	13
Beta	b1	Closing feed inventory	10	4	4	0	4	4	4	5	5	4	4	4	4	4
	a2	WIP inventory	2	1	0	0	0	0	0	1	6	4	2	2	1	1
		Finished goods inventory	2	1	0	1	0	0	0	0	0	0	0	0	0	0
			Total average daily units =													**37**
			At an average despatch rate of 198 units/day of Alpha + Beta, the Days of Inventory (DOI) =													**0.19**

Table 5.10 Source, Make and Deliver Sub-Processes of Order Fulfilment

Level Three	*Level Four*
Source	Input is sales and operations plan Confirm supplier capacity availability Confirm supplier lead times Determine optimal supplier selection Determine order amounts and timings Place orders on suppliers Track supplier order status Track supplier order DIFOTEF results Review sourcing and improve
Make	Input is sales and operations plan, MPS and Unit Schedule For each time period, launch necessary feed into SC Ensure constraint resources are fully utilised and buffered Eliminate all waste and ensure all 'make' costs are minimised Check that scheduling rules are observed Check that all operating units are observing the unit schedule Ensure error and defect free product/service quality Continuously work to reduce all set-up times Continuously work to improve constraint unit throughput
Deliver	Input is despatch plan from S&OP and MPS Confirm order is ready for shipment Confirm customer is ready to accept shipment and specifically when and where Confirm transport arrangements Confirm any necessary warehouse space Despatch the order Track the order Provide delivery confirmation to customer and to finance department Issue invoice for payment (if applicable)

Table 5.11 The Seven Managing Returns Sub-Processes

Level Three	*Level Four*	*Level Five*
Returns request notification and initial response	Return notification with details. Using a digital notification system, a downstream SC partner/ customer notifies that a return is warranted	Product order details, including specific return units order item numbers Reason(s) for return items Date reason(s) detected Current location of the affected items Key contact personnel Inspection arrangements Inspection execution Initial response to return notification
Return routing	Digital expert system determines optimal return routing path	Launch optimal returns routing system
	Return transport details	Log all return details
	Collection timing	Launch pick-up process
Receive return	Notify customer of return receival	Email, SMS, CRM system, customer portal
	Track time duration from initial notification to return receival	Input to plan, do, check, act (PDCA) improvement cycle

Level Three	*Level Four*	*Level Five*
	Track duration from time of return receival to customer advisal of receival	- ditto -
	Log return receipt into returns management system, including return details, return reason(s), storage location and timings	- *ditto* -
Return treatment decision	Specialist return treatment staff undertake inspection and determine appropriate return treatment	Repackage Repair Refurbish Remanufacture Scrap the product and make replacement or issue credit note Notify relevant returns processing teams
Return processing	Process the return as per advised return treatment Log processing results	Launch return processing actions
Continuous improvement	Apply PDCA improvement cycle to all returns activities Publish results of continuous improvement activities as a 'Quality Story' Recognise improvement teams for all positive results achieved	
Return costs	Produce return performance report, including recommended improvement actions	Return transportation costs Return treatment decision costs Return processing costs Continuous improvement team costs Return support system costs Potential sales loss from return arisings

their deployment. Ongoing monitoring, review and continuous improvement of each process is also a priority.

The fifth level two SC *execution* sub-process is that of *enabling suppliers*. This sub-process has nine sub-processes; the first four of which are shown in Table 5.12. Again, each of the level four and level five sub-process identified need to be defined for the particular SC involved and then actualised.

The last five *enabling suppliers* sub-processes are shown in Table 5.13. Conducted competently and comprehensively, these sub-processes contribute significantly to the performance of all SCs.

The sixth and involved SC execution sub-process is that of *collaboration, cooperation and integration*. This sub-process has four level three sub-processes; the first of which is SC *relationship management* as shown in Table 5.14.

The second sub-process of *collaboration, cooperation and integration* is that of SC *information sharing* as shown in Table 5.15.

Table 5.12 First Four Enabling Suppliers Sub-Processes

Level Three	*Level Four*	*Level Five*
Categorise supply base	Strategic suppliers (number and identity)	Collect data, cleanse and verify
	Preferred suppliers (number and identity)	Pass suppliers through 'selection sieve'
	Transactional suppliers (number and identity)	*- ditto -*
	Determine desired mix of previously mentioned categories	Confirm and seek approval
	Take action to deploy desired mix	Implement approved decision(s)
Make, buy, outsource decisions	Product and service requirements	Collect data, cleanse and verify
	Supplier capabilities	*- ditto -*
	Supplier accreditations	*- ditto -*
	Total cost, environmental and social comparison	Make analysis Buy analysis Outsource analysis
	Full risk analysis – identification, rating, ranking and response determination for all risks	Conduct formal risk analysis (see Chapter 3)
	Seek approval for the recommended option	Make a recommendation based on the earlier analysis
Supplier appraisal	Quality of leadership Market position Life-cycle costs Technical capabilities Social capabilities Key KPI performance Attractiveness of products and services	Request data from suppliers + collect data from other sources, cleanse, verify and analyse
Rationalisation of the supply base	Rating of each supplier based on level of investment matrix	Organisation's level of investment Suppliers' level of investment
	Decision of supply base make-up based on investment matrix analysis	

Table 5.13 The Final Five Enabling Suppliers Sub-Processes

Level Three	*Level Four*
Developing supplier potential	Involving suppliers in organisation's strategic planning process
	Supplier involvement in new product developments
	Encouraging supplier involvement in technical agreements for adoption of new process practices and/or technologies
	Provision of key personnel to suppliers for the adoption of new methodologies
Early supplier involvement in new product development	Jointly establishing customer needs, wants and desires
	New product design and specification development
	Matching supplier requirements schedule with new product launch schedule
Supplier negotiations	Agreement on common goals
	Agreement on key information sharing
	Agreement on guiding coalition
	Agreed negotiation process and schedule
	Agreement on profit sharing formula

Level Three	*Level Four*
Supplier relationships (including partnerships)	Establish and actively manage formal supplier relationship agreements
Managing supplier performance	Set clear supplier performance expectations Clearly identify the consequences of positive and negative performance results Monitor and publish individual supplier performance results on key KPIs Make available continuous improvement specialists to suppliers at no cost, and encourage suppliers to make use of them on jointly agreed improvement initiatives Hold regular reviews, assess performance, recognise positive performance, agree on any necessary corrective actions and/or new joint initiatives

Table 5.14 The First Collaboration, Cooperation and Integration Sub-Process – SC Relationship Management

Level Three	*Level Four*	*Level Five*
SC relationship management	Develop joint 'Relationship Agreement' with SC partners involved	Agree on the specific *purpose* of the relationship Agree on joint SC goals Agree on relationship ground rules and expected behaviours Agree on the principle of collaboration Agree on the principle of information sharing Agree on the principle of SC personnel, process and systems integration Agree on the communications process to be followed Agree on improvement results sharing
	Develop and agree on specific collaboration initiatives	Engage with SC partners, identify SC beneficial initiatives, set resourcing and timings
	Implement agreed collaboration initiatives	Give clear signal to act, review and correct as necessary
	Implement agreed communication process	Develop multi-channel and multi-issue frequency communication methods
	Monitor ongoing performance	Identify issues and challenges Identify performance gaps and take corrective action Publish regular performance reports, including results of all corrective actions
	Periodic formal review and agreed improvement actions	Schedule audits and reviews into the SC management system

Table 5.15 The Second Collaboration, Cooperation and Integration Sub-Process – SC Information Sharing

Level Three	*Level Four*	*Level Five*
SC information sharing	Identify all information which, if shared, would improve SC performance	Conduct search and analyse data availability and attractiveness of sharing discovered data and information
	Determine and agree on how to share such information	List of information elements identifying both width and depth How shared? (i.e. access details) With whom? (including security requirements) When?
	Specify how the SC information is to be used to improve all SC material flows, inventories and customer service performance	Provide examples on how the data and information can be used for improvement purposes
	Share relevant information with customers	Detailed product/service offer (e.g. new products, offers, product/service catalogue and updates, prices and payment terms) How to order and order placement Details on any customer credits Invoices Order status New product/service introduction details
	Share relevant information with SC partners	Business/market intelligence Economic development knowledge Demand levels per SKU (e.g. from POS data) SC unit capacities per SKU SC plans and schedules SC inventories per inventory type (RM, WIP, FG) per SKU per stocking location Safety stock status per SKU per safety stocking point SC product path lead times New product introduction schedules SC constraint resource status Product/service quality performance Scheduling rules Set-up times SC ABC cost details Results of SC optimisation analytics Relevant details on SC payments Details on all SC returns

The last two *collaboration, cooperation and integration* sub-processes are those of *SC performance measures* and *SC people effectiveness execution*. The level four and level five sub-processes involved with each of these are shown in Table 5.16. Data on each of the performance measures needs to be collected, cleansed, analysed and reported. Corrective actions need to be taken for all performance measure shortfalls. Execution of the SC people effectiveness sub-processes provides the human capability to affect all of the SC *execution* sub-processes.

Table 5.16 The Final Two Collaboration, Cooperation and Integration Sub-Processes – SC Performance Measures and SC People Effectiveness Execution

Level Three	*Level Four*	*Level Five*
SC performance measures	% DIFOTEF	Delivered in-full, on-time and error free
	Order cycle time (order lead time)	Duration from customer order placement to delivery
	Responsiveness to changing customer demand levels	Time taken to respond to demand increase or decrease
	Flexibility to changing customer requirements	Ability to accommodate changes to order amounts, due dates, product types, product characteristics
	Resilience to SC disruptions	Ability to withstand shocks (see SC resilience builder in Figure 2.8)
	SC ABC costs	Activity-based costs are essential for many SC optimisation calculations
	Days of inventory	Volume of all inventories divided by daily despatch rate
SC people effectiveness execution	Deploy organisational structural design	Accountability hierarchy
	Deploy expectations	One position description for each approved role
		Clear job goals
		Clear SC goals
		Clear responsibilities
		Clear authorities
	Implement accredited training and assessment process	No 'sign-on' or access authorities unless training finished and assessment passed
	Formation of required teams	Careful selection of team leader and team members
	Establishment of 'Communities of Practice' (COPs)	Define highest value-add COPs and establish them
	Provision of all necessary resources	Sufficient and competent personnel
		Appropriate tools
		Necessary equipment
		Useful techniques
		Well-written and illustrated operating manuals
		Complete and up-to-date set of Standard Operating Procedures (SOPs)
	Definite signal to perform	Individuals, teams and communities of practice
	On-the-ground leadership support, presence and encouragement	Apply leadership processes explained in detail in Chapter 6
	Execution of designed recognition and rewards system	- *ditto* -
	Implementation of the effectiveness appraisal process	- *ditto* -

The last SC *execution* sub-process is that of *support systems*, specifically ICT support systems. As SCs become ever more digital, the professional execution of the sub-processes involved here is vital to modern-day SC competitiveness. This level two sub-process has six sub-processes; the first three of which are shown in Table 5.17.

The last three *support systems* sub-processes are *e-Commerce execution, process integration execution* and *ICT project management*. The level four and level five sub-processes for each

Table 5.17 The First Three Support Systems Sub-Processes – ICT Plan Execution, ICT Infrastructure Execution and Software and Applications Execution

Level Three	*Level Four*	*Level Five*
ICT plan execution	Strategic SC ICT initiatives	New product to market lead-time reduction Total *Customer Care* system (CRM) Reliable, easy to use, error checking order capturing system End-to-end SC visibility system SC cycle time reductions Suppliers/Service providers integration (SRM) Collaboration projects facilitation SC resilience building SC personnel productivity improvements SC cost reductions
	ICT technology adoption plan execution	Formal project readiness & assurance process
	ICT disaster response and recovery plan execution	- *ditto* -
	ICT security policy and plan execution	- *ditto* -
	ICT annual budget	Data collection, cleansing & analysis for budget formulation
ICT infrastructure execution	On-Premises Computing (input from SC *design*)	Facilities plan implementation Hardware plan implementation Software plan implementation Networks plan implementation Protocols plan implementation Staff plan implementation
	Cloud (input from SC *design*)	Structure plan implementation Service model plan implementation Scope plan implementation Data security plan implementation Staff plan implementation
Software and applications execution	Customer Relationship Management system (CRM) Supplier Relationship Management system (SRM) Planning & scheduling Manufacturing Execution System (MES) SC operator assistance systems Transport Management System (TMS) Warehouse Management System (WMS) Returns Management System (RMS) SC analytics system	Implementation of approved designs using 'project readiness and assurance' process

of these are shown in Table 5.18. Again, these three SC sub-processes are key to SC attractiveness and performance. The included electronic total customer care system for example, if implemented, would be a key marketplace differentiator.

Table 5.18 The Final Three Support Systems Sub-Processes – e-Commerce Execution, Process Integration Execution and ICT Project Management

Level Three	*Level Four*	*Level Five*
e-Commerce execution	Electronic Total Customer Care (TCC) system	Product/service catalogue Customer enquiries system Customer ordering and payment system Customer order status reporting Order delivery confirmation system Customer rating of delivered product/service system On-line manuals and help system Customer community system e-Payment system and records Claims and returns system
	Process integration for all supplier and service provider ordering and payment	
	SC visibility system for all relevant SC data and information	
	Supplier claims and returns system	
Process integration execution	Ratification of agreed objectives and goals	
	Objective confirmation of all specific process integration opportunities that genuinely add value	
	Detailed description of all integration risks and challenges	Economic, environmental & social
	Specific project business cases and approvals	
	Implementation of approved projects	Project management Project readiness and assurance process
	Implementation of designed review and improvement process	
ICT project management	Phasing of all ICT projects	Resource availability, aimed completion times, project priorities
	Implementation schedule for all approved projects	What, who, when, budget, constraints
	Business case development schedule for all intended but non-approved ICT projects	Future projects schedule, including estimated cost and benefits
	Execution project team management	Project leader selection execution Project member selection execution Project team members performance review process execution Project team member re-entry process execution Project team register Identification and management of all team members who are resource constraints
	Project readiness and assurance process activation	
	Enacting approved project management review and improvement process	

To finish this section, a detailed sub-process example is now presented, being that of *project readiness and assurance*. This sub-process is essential to the successful delivery of all projects of course but is also very relevant to any projects undertaken on SC *execution* processes, especially the ICT projects underpinning delivery of a fully digital SC.

The aim of *project readiness and assurance* is to ensure that projects under consideration are implemented on-time, on-budget and to-promise. To-promise means that all promised functionality is actually delivered.

Like all of the processes described in this book, *project readiness and assurance* is made up of a series of sub-processes (sometimes called project stages) as shown in Figure 5.2. It is important that each stage of the project is conducted by competent and experienced personnel and that *each stage* is subject to project assurance assessment.

Project *assurance* is a process which, if enacted diligently, can indicate to SC leaders that the project will be capable of delivering all of the expectations that have been set for the project when the project is completed. Again, project assurance needs to be carried out by subject matter experts who are separate to the main project team. Such people are not acting in a police type role, rather they are competent and unbiased auditors who ensure that all necessary project conditions are met at *each stage* of the project.

The 'STOP' signs evident in Figure 5.2 are what some project professionals refer to as 'stage gates'. That is, the gate to the next stage doesn't open unless all the conditions for the current stage and the planning requirements for the next stage are met. This is really the assurance test. If the project passes the gate test, then SC leaders can be assured that the project is on track and will be successful. Such conditions vary somewhat from stage gate to stage gate; however, common stage gate conditions are:

i Is this project aligned with the overall SC strategy?

ii Has the target market been clearly defined and have customers within that market been tested as to their acceptability of the initiative?

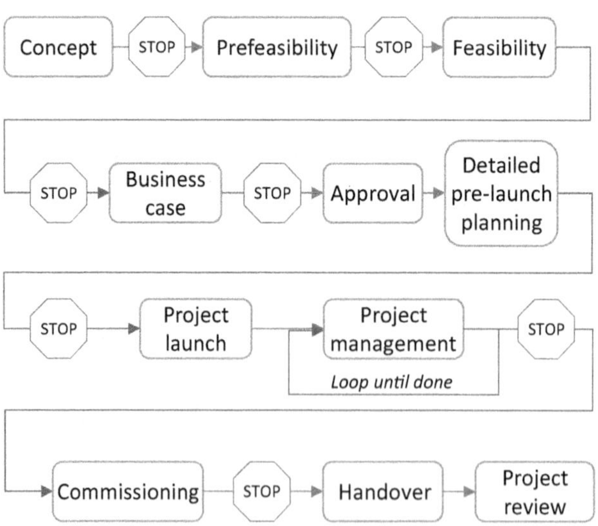

Figure 5.2 Stages of the *Project Readiness and Assurance* Process

iii For the stage just finished, does demonstrable evidence exist that all of the required steps have been properly completed?
iv For the stage just completed and for the next stage, is the data that has been collected and that has been used for design and decision-making purposes relevant to the task and has it been cleansed and verified?
v Was the data analysis carried out, conducted to reliable data analysis standards using tested data analysis techniques?
vi What was the on-time performance of the activities for the stage just completed?
vii What was the spend rate against the spend rate budget for the stage just completed?
viii Has a formal stage report for the stage just completed been compiled and circulated to all relevant personnel?
ix What is the probability the next stage's chosen option will deliver its defined expectations?
x Is the next stage's chosen option a proven one? Has it been tested in practice at full scale?
xi What are all of the risks involved with the next stage's proposal? How are such risks to be actively managed?
xii Has a detailed and professional project plan, detailed budget and spend rate forecast been prepared for the next project stage?
xiii What is the order of accuracy of predicted project outcomes for the next stage?

Typically, one of three outcomes are possible after a stage gate assessment:

i A *red* light – Meaning it is not 'safe' environmentally and/or socially and/or economically to proceed, so stop
ii An *orange* light – Meaning important activities from a previous stage are incomplete or that things are changing, so further work and/or further study is required before proceeding
iii A *green* light – Meaning it is 'safe' to continue, so proceed to the next project stage

The individual process steps shown in Figure 5.2 can be summarised as:

i *Concept Stage* – This stage involves ideas generation and assessment. For example, consider the case whereby a SC professional believes it to be a good idea if their SC could develop and implement a relevant *total customer care* digital system. The idea is to create a seamless customer experience by making all customer interactions electronic and automated, using artificial intelligence as much as possible. The concept stage for this idea therefore would include a detailed description of the project outlining its vision, purpose, objectives, scope, target market, level of customer want/need/desire, a description of how the system would work, a listing of all benefits including sustainability benefits, a listing of the expected scale of the benefits, estimated project cost, project timeline and, importantly, a full analysis of all risks. The order of accuracy of benefits and costs for this stage is typically +/- 35%. All of the previously mentioned considerations need to be captured in a formal stage report.
ii *Prefeasibility Stage* – Many of this stage's steps are similar to those of the *concept* stage albeit conducted in more detail. For example, the safety, environmental, social and economic considerations all need to be expanded on including the calculation of the project's expected Net Present Value (NPV). There are three important additions

necessary for this stage however: the first is that the *prefeasibility* stage needs to include *options generation* and *options assessment* in order to select the preferred one. The second concerns defining detailed *risk management strategies* for all risks identified during the *concept* stage. The third addition is that of describing the *key outcome expectations* for the *feasibility* stage. Preliminary *pre-launch planning* also starts in this stage. The order of accuracy of benefits and costs for this stage is typically +/- 25%. As per the *concept* stage, all of the previously mentioned considerations need to be captured in a formal *prefeasibility stage* report.

iii *Feasibility Stage* – Again, many of the steps involved here are the same as the first two stages albeit in much more detail. Importantly for this stage, a number of project factors need to be finalised, including confirmation of the preferred option, the scope, risk analysis, time schedule, resourcing requirements, expected benefits, full project cost and expected expenditure timing. The stage team also needs to confirm that the project's returns meet the necessary SC performance hurdle rates. The order of accuracy of benefits and costs for this stage is typically +/- 15%. *Pre-launch planning* is also further progressed in this stage Again, all of the previously mentioned considerations need to be captured in a formal *feasibility stage* report.

iv *Business Case* – This stage is primarily about making the case for project approval, including, importantly, funding approval. The *business case* therefore needs to be captured as a formal approval-submission document. It is the document sent to key decision makers in advance of the submission presentation. As such, the *business case* needs to include a succinct summary of all of the relevant descriptions, analyses, results, conclusions and forecast outcomes contained in the *feasibility* report. The *business case* is about making the case for the project: Why should the SC invest time, resources and money in this project? Who are the beneficiaries? Do customers want this? Who exactly are those customers? What are all of the benefits? Who gets to share the benefits and how? Does this project provide the SC with any competitive advantage(s)? What are all of the risks and the risk management strategies? What are the sustainability implications? What are the economic and non-economic returns and what is the level of confidence that those claimed returns will actually be delivered? What is the order of accuracy of the estimates contained in the *business case*? What is the project timing? How does this project synchronise with other projects the SC has approved for progression? The order of accuracy of business case submissions is typically +/- 7.5%.

v *Detailed Pre-Launch Planning* – This stage is about developing a detailed project execution plan. This plan builds on the *pre-launch planning* work undertaken during the *prefeasibility* and *feasibility* stages. The detailed project execution plan comprises plans for each of safety, environment, social (including community), overall project management (what, who, when, duration, sequence, constraints, cost), budget and spend rate, spend approval process, engineering, quality, permitting and approvals, construction, contracts, supply, logistics, human resources, risk management, access and security, commissioning, data and information capture, analysis and dissemination, performance management (regular reviews and reports, project ratings and corrective actions).

vi *Project Launch* – This is about giving a formal signal to proceed with the project's execution. It involves the actualisation of the detailed plans finalised in the previous stage. Normally, this signal to start execution is given by senior personnel at a project execution 'kick-off' session. The project team and all key stakeholders should attend this session where the overall project plan will be presented, the project management methodology to be adopted clarified, deliverable expectations of the project team reinforced, authorities and responsibilities clearly spelt out and a genuine expression of confidence given.

vii *Project Management* – This is an iterative process that includes regular reviews of the project plan itself, project progress to time plan, resource management including identification of all constraints and active constraint management, project funds spend rate, the crafting and launch of necessary corrective actions, formal project review reports and dissemination. An important and often overlooked process here is that of 'health checks' of project team members in case any member is approaching an overload or even a 'burn out' condition.

viii *Commissioning* – As project execution nears its conclusion, the commissioning of various parts of the project can occur in parallel. Project execution really ends when all project components are satisfactorily commissioned. Commissioning involves the safe powering up of electrical equipment, checking of valve and actuator operation and fluid and gas flows, instrumentation, control loops and software, process stationary testing, process dynamic (running) testing, performance checking and corrective actions. A formal *commissioning report* needs to be completed for each and every project item commissioned. Operational personnel need to sign-off on each element of this report before project personnel are relieved of their responsibilities.

ix *Handover* – This is the point where, having signed-off on the commissioning reports (meaning satisfaction with the commissioning work), the operational personnel formally take responsibility for the ongoing operation and maintenance of the new asset. This *handover* stage is best recognised with a formal ceremony, both to thank and recognise the project team and express confidence in the operational team who now 'own' and are responsible for the asset. Very importantly, *all* of the project team members need to be carefully and promptly re-entered to their previous roles or reassigned to new roles. The worse thing that can happen to project team members at the end of a project is to be left 'hanging' with no position to go to. This is very bad for the project team members directly involved of course, but it also discourages personnel from participating in any future projects.

5.4 Chapter 5 case study – achieving supply chain *execution* excellence in practice

The case study for this chapter involves the top three rated companies in Gartner's 'Supply Chain Top 25 for 2020' (Gartner, 2020). The Gartner process assesses large (> USD 12 billion of revenue) manufacturing, retail and distribution organisations.

The top three companies rated for their SC environmental, social and economic excellence in the 2020 list were:

i Cisco Systems
ii Colgate-Palmolive
iii Johnson & Johnson

Taking each of these companies in turn, the reasons for their top ratings can be summarised as:

Cisco Systems – This hi-tech company scored highly on community opinion, demonstrable focus on environmental, social and governance performance including embracing the concept of the circular economy (recycling, repair and reuse strategy on all materials used and product supplied), customer focus with an offer-based value proposition, development of a fully digital SC with built-in security features, product and service lead-time reliability, inventory and cost reductions, all whilst successfully bringing many new products and services to market.

Colgate-Palmolive – A manufacturer and supplier of consumer products, this organisation has a very strong ecological record and received certification as a 'TRUE Zero Waste' organisation from the US Green Building Council. Colgate-Palmolive's aim is to have 100% of its products recyclable by 2025. To this end, Colgate-Palmolive has already launched a fully recyclable tube and has shared the technology with other companies such that all tube products may be recycled.

Johnson & Johnson – A long established healthcare products supplier, Johnson & Johnson has embraced SC innovation as a key marketplace differentiator. The company has developed a facility in New Jersey, for example, that facilitates the collaboration of Johnson & Johnson's SC partners with external specialists in the development and testing of new ideas to improve healthcare. A practical example of useful ideas being actualised is that of the application of Johnson & Johnson's 3D printing expertise to 3D print new manifolds for ventilators used on COVID-19 infected patients. During the 2020 pandemic, ventilators were in short supply, so the 3D printed manifold allowed two patients to share one ventilator.

It is interesting to note that none of the 2020 rated top three were part of Gartner's 2020 'Masters' group. The 'Masters' classification was introduced by Gartner in 2015 and includes companies that achieved top five composite scores for seven of the previous ten years. The 2020 Masters group was made up of Amazon, Apple, McDonald's, Proctor and Gamble and Unilever.

In its 2020 report, Gartner noted that individual Masters' group members were focused on one or more of the following:

- i Sustainability – Especially issues concerning recycling, reuse and environmental footprint impact
- ii Digitisation of the SC and especially increased end-to-end visibility
- iii Distribution capacity and cycle time reductions as on-line ordering ramps up
- iv Response to SC shocks (especially the 2020 COVID-19 crisis)
- v Enhanced customer experiences
- vi Cost management

As a result of the Gartner 2020 ratings and results analysis, the following recommendations are posited for operating SCs to consider:

- i The development of a clear, inspirational and modern *purpose* that embraces all of the components of sustainability (i.e. environmental, social and economic) and not just the economic component alone
- ii Build SC *nimbleness* and *robustness* so as to be able to compete and survive in today's volatile and disruptive marketplace
- iii Build SC digital competence and expertise. Use this to coordinate all of the SC digital elements into a SC end-to-end electronic control system that enables SC collaboration and underpins SC *execution* excellence

5.5 Review questions

- i Why is SC *execution* performance of such importance to customer relationships?
- ii What metrics can be used to determine if the *new products and services delivery to customers* process is working effectively?
- iii Would *sales and operations planning* be considered as a critical SC *execution* process? If yes, then why so?
- iv Do you consider that there is a high or low level of interrelationship between *sales and operations planning* parameters? Give reasons for your answer
- v How might facility utilisation rates (loading levels) impact SC resilience? How should such impacts be managed?
- vi Why are the SC *support systems* processes so important to modern-day SCs? How might SCs leverage this importance?
- vii What are the common traits of the top three companies in Gartner's 2020 top SCs list?

5.6 Assignment topics

(Each to be 3,000 words minimum or about three A4 pages single spaced. Tables and figures are not to be included in the word count.)

- i Using a pseudo-algorithm style, develop an algorithm that delivers a workable sales and operations planning process. Use the sales and operations plan input conditions and outputs as shown in Table 5.7 as the basis of your algorithm. Code your completed algorithm into a Microsoft Excel spreadsheet to demonstrate its workability
- ii Provide a written description of each of the SC *planning* sub-processes. Include diagrams to illustrate the relationship of each sub-process to the others, including where they exist with one another over a time horizon
- iii Using your own research, provide a detail description of just why it is that Amazon, Apple, McDonald's, Proctor and Gamble and Unilever have appeared on Gartner's 'Masters' list for so many years
- iv Using your own research, describe in detail how a competent *project readiness and assurance* process works. Discuss the applicability of such a process to the management of SC *execution support systems* (e.g. digital) projects

5.7 Chapter 5 reference

Gartner. (2020). The Gartner Supply Chain Top 25 for 2020. Available at: https://emtemp.gcom.cloud/ngw/globalassets/en/supply-chain/documents/trends-top-25/gartner-supply-chain-top-25-2020-white-paper.pdf (Accessed 26 August 2020).

6 Recruiting, developing and enabling supply chain *people*

6.1 What you will learn in this chapter

The chapter starts with an introduction to the four SC *people* processes and the key metrics that can be used to assess their performance.

Next, the levels four, five and six sub-processes of each of the six level three *people* sub-processes are defined. In this section, three detailed examples are described: a *leadership suitability* process, a *position description* process and an *effectiveness appraisal* process.

In addition to the complexity of the SC *people* processes themselves, there are also a number of obstacles and barriers that SC professionals will encounter in trying to implement them. Such obstacles and barriers are identified, including suggested responses to overcome them.

The chapter's case study involves a research study undertaken on 59 companies operating within SCs in New Zealand (NZ). The researchers successfully identified five barriers to and four enablers of effective SCM and the development of a working SC *culture*. In particular, the researchers made the point that people management issues are *the* most important in the achievement of SC excellence.

The chapter concludes with a relevant set of review questions and written assignment topics.

6.2 Supply chain *people* sub-processes and metrics

SC *people* processes impact all other SC processes whereby the quality of SC people and the quality of the processes they use greatly influence SC outcomes. Overall SC *excellence* really starts with SC *people* excellence. And the effect is iterative because SC *people process* excellence drives SC *people* excellence which in turn drives SC process excellence. This dynamic is illustrated in Figure 6.1.

Figure 6.1 SC Process Relationships

The importance of quality SC people using quality SC processes therefore cannot be emphasised enough. Organisations that understand such importance take the matter of people and the processes they use very seriously, including investing significantly in the development of both. Example organisations include Cisco Systems, Google, IAG, Intuit, Pixar and Unilever.

The four level two SC *people* processes considered in this chapter are:

i SC leadership
ii SC managerial system
iii SC culture
iv SC knowledge building

As for the first three SC processes of *strategy*, *design* and *execution*, the four SC *people* processes are all highly interrelated. The development and continuous improvement of the *people* processes therefore needs to be inclusive of all of them, and any changes made to one need to be checked for the impact such changes have on the other three.

The first two level two processes along with the corresponding level three sub-processes and level three key metrics are shown in Table 6.1.

The last two level two processes along with the corresponding level three sub-processes and level three key metrics are shown in Table 6.2.

Some SC leaders focus their attention predominately on the first two SC *people* processes: SC leadership and the SC managerial system. Such focus, however, ignores the reality that the SC people processes are really a *gestalt*; that is, SC leadership and managerial

Table 6.1 First Two Level Two SC People Processes Plus Level Three Sub-Process and Metrics

Level Two	*Level Three*	*Metrics Used for Level Three*
SC Leadership	Setting direction	A published, approved, up-to-date and comprehensive business strategy exists
	Leveraging the potential of all resources	Personnel, physical assets and non-physical assets effectiveness indexes
	Leader suitability	Leader technical suitability index Leader social suitability index
	Leader performance	Leader effectiveness index
SC Managerial System	Requisite-organisation	Approved organisational design that is aligned with organisational strategies exists Approved written position descriptions for each position exist Approved remuneration policy exists Active effectiveness appraisal process exists Approved code of conduct exists Personnel inflow/outflow management system exists Selection, training and assessment systems exist ICT support systems requirements definition exists
	SC people costs	Total personnel costs Total costs per role type Costs per hour overall Costs per hour per role type Productivity index

Table 6.2 First Two Level Two SC People Processes Plus Level Three Sub-Process and Metrics

Level Two	*Level Three*	*Metrics Used for Level Three*
SC Culture	Principles, values and behaviours	Safety performance Number of appropriate behaviours per employee Number of inappropriate behaviours per employee Personnel turn
SC Knowledge Building	Open exploration and sharing of ideas	Active communities of practice exist, active recognition of best practice exists, technical support systems exist
	Collaborative learning	Number of active departmental, cross-functional improvement teams, number of active supply chain, cross-functional improvement teams, plus ratios of these to number of employees
	Advancing knowledge	Approved knowledge capture process exists, knowledge storage system exists, knowledge analysis and interrogation system exists, password protected, tiered access exists for all SC personnel

systems simply cannot be effective if the SC culture is wrong and/or knowledge building is inactive along with weak corporate memory.

6.3 Supply chain *people* processes in more detail

Prima facie, the four level two SC *people* processes seem rather straightforward; that is, there are only four of them. How can that be hard to manage? The reality is rather different, as will be appreciated by studying the levels three, four, five and six sub-processes that make up the set.

The first level three sub-process of *setting direction, leveraging the potential of all SC resources* and *SC leader suitability* along with their corresponding levels four, five and six sub-processes are shown in Tables 6.3a, 6.3b and 6.3c.

At this point, a detailed process example will be presented, being that of SC *leader suitability*. The first issues to consider with this process are those of leader *technical* suitability and leader *social* suitability. Tables 6.4 and 6.5 capture each of these requirements, including the specific criteria and the minimum and preferred level on each for the example position of SC leader.

The suitability criteria are integral to the suitability assessment process used for new hires and the suitability assessment used as part of an overall effectiveness appraisal process. Each of these processes are illustrated at Figures 6.2 and 6.3 respectively. As can be appreciated, it takes considerable time and effort to execute these processes well. However, the pay-off, so long as the processes are executed competently, is considerable and certainly worth the effort.

Table 6.3a Setting Direction Sub-Processes

Level Three	*Level Four*	*Level Five*	*Level Six*
Setting direction	Developing an aspirational SC visioning process	Identification of vision elements, element target setting, vision refining	Vision development schedule, participants, process steps, including active engagement & involvement process + approval
	Selling the vision process	Communication engagement and marketing circular process	Time schedule, identification of target audience, personnel involvement & feedback capturing
	Setting goals and targets process	Identification of key outcome measures, comparative performance assessment, target setting	Goals development schedule, participants list, process steps, including active engagement & involvement process + approval
	Jointly developing effective and aligned strategies	Definition of key actions needed to deliver vision element targets	Strategies development schedule, participants list, process steps, including active engagement & involvement process + approval
	Risk management process	Identification of all risks, risk rating, risk categorisation, description of preventative & contingent actions	Risk appraisal schedule, participants list, process steps, including active engagement & involvement process

Table 6.3b Leveraging the Potential of All SC Resources Sub-Processes

Level Three	*Level Four*	*Level Five*	*Level Six*
Leveraging the potential of all SC resources	Process for the selection of high calibre staff	Matching potential hires to position requirements	Position description process Detailed hires recruitment, assessment and selection process
	Building relationships process	Definition of relationship targets, description of desired relationship types and levels, description of how it is to be done	*See Table 5.14*
	Ensuring strategy implementation process	Project readiness & assurance process	*See Section 5.3*
	Coaching, encouraging and supporting process	Identification of coaches and people to be coached, definition of outcomes and coaching regime to be followed	Goals, actions, implementation, measures, reviews and recognition
	Auditing, effectiveness reviews and improvement process	Definition of audit/ effectiveness review scope, targets, expected outcomes and process steps to be followed	Scheduling, execution, results analysis, reports, corrective actions
	Monitoring of results achieved against goals process	Definition of results to be monitored, data capture process, analysis and reporting	Schedule of activities (process steps), identification of who is responsible and timing required

Table 6.3c SC Leader Suitability Sub-Processes

Level Three	*Level Four*	*Level Five*	*Level Six*
SC leader suitability	Technical suitability of SC leaders' assessment and building process	Qualifications and accreditations	Tertiary institutions, professional bodies, accreditations, certifications (e.g. Six Sigma black belt)
		Experiences	Industry, position, duration
		Specialities	E.g. Six Sigma, Lean, Theory of Constraints, digital supply chains
		Specific talents	E.g. design, mathematics, spatial ability, innovativeness, agility
	Social suitability of SC leaders' assessment and building process	Stability	Assessment, coaching & improvement process
		Maturity	*- ditto -*
		Sobriety	*- ditto -*
		Responsibility	*- ditto -*
		Trustworthiness (honesty, self-respect, respect for others, listening skills)	*- ditto -*
		Openness and inclusiveness	*- ditto -*
		Effective communicator	*- ditto -*
		Perseverance	*- ditto -*
		Determination	*- ditto -*
		Availability (is present and accessible)	*- ditto -*

Table 6.4 Leader *Technical* Suitability Criteria and Targets

Position	*Supply Chain Leader*		
	Criteria	*Minimum*	*Preferrable*
Technical Suitability Standard	Qualifications & accreditations	BBus (SCM) or BCom (SCM)	MSc (SCM), ALC, ASCM, CIPS or CSCMP accreditation
	Experiences	5 years minimum in operating SC professional role with demonstrable performance results	> 10 years of uninterrupted SCM experience in > 2 organisations and with demonstrable performance results
	Specialities	Proficient in all aspects of Lean and SCM	Theory of constraints accredited, Six Sigma black belt, application of fully digital SCs
	Specific talents	SC process expertise	SC analytics, design skills, mathematics proficiency

Table 6.5 Leader *Social* Suitability Criteria and Targets

Position	*Supply Chain Leader*		
	Criteria	*Minimum*	*Desirable*
Social Suitability Standard	Stability Maturity	Top quartile on personality test	Top decile on personality test
	Sobriety	All-clear assessment made by an addiction professional	All-clear assessment made by an addiction professional
	Responsibility Trustworthiness Openness and inclusiveness Effective communicator Perseverance Determination	Top quartile on personality test	Top decile on personality test
	Availability (is present and accessible)	Favourable rating on most recent effectiveness review	Superior rating on most recent effectiveness review

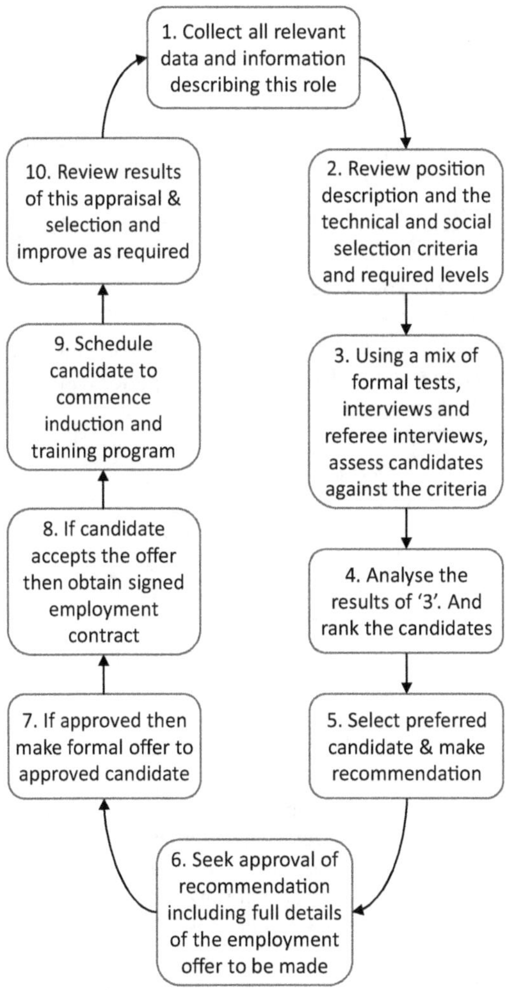

Figure 6.2 Suitability Assessment Process for New Hires

The levels four and five SC *requisite-organisation* sub-processes, as shown in Table 6.6a and 6.6b, are very good examples of just how involved and lengthy the SC *people* processes are.

A second process example will now be illustrated by way of a *position description* for the position of SC Planner.

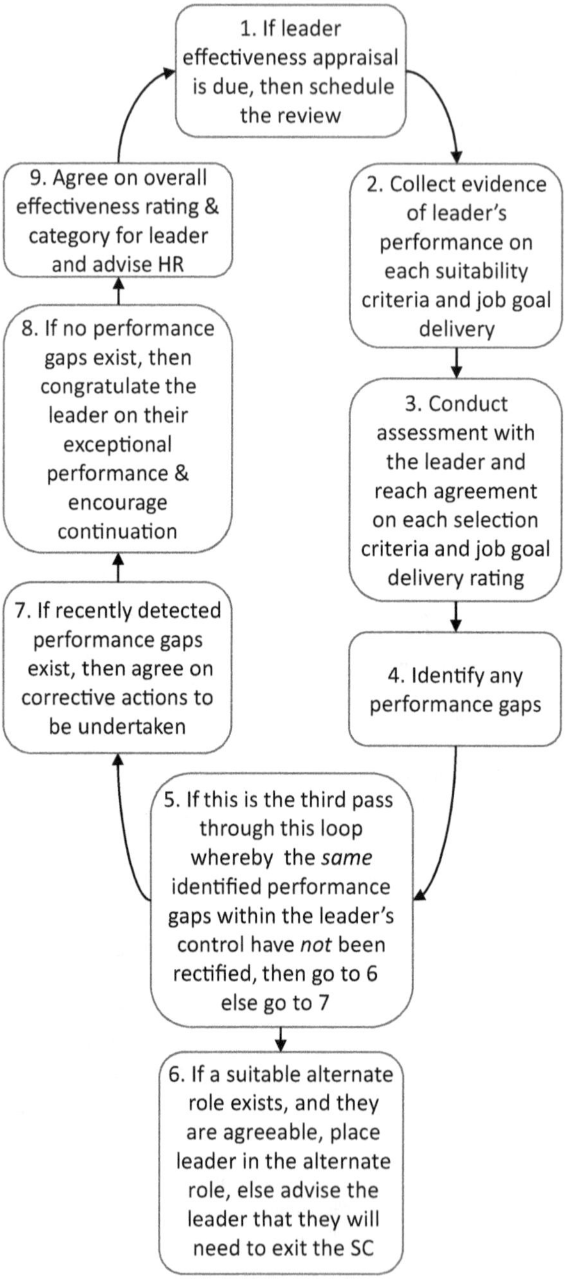

Figure 6.3 Ongoing Suitability Assessment Process as Part of the Effectiveness Appraisal Process

Table 6.6a The First Four SC Requisite-Organisation Sub-Processes

Level Three	*Level Four*	*Level Five*
SC requisite-organisation process	Organisational design process	Procedure for the mapping of roles against processes
		Process for the calculation of the numbers of personnel required per role and in total
		Process for assessing alignment with organisation's strategies
		Process for setting accountability and authority hierarchy
		Functional stewardship definition
		RACI matrix development process
	Compilation of position descriptions process	Purpose of role guidelines
		Setting of specific job goals guidelines
		Guideline for setting clear accountabilities and authorities
		Process for determining skills and capabilities required
		Process for determining training and assessment requirements
		Process for determining tools and equipment needed
		Process for determining relationships with adjacent roles
		Process for setting effectiveness appraisal schedule
		Clear positive and negative consequences statement
	Remuneration process	Base salary/wage determination process
		Short-term incentives setting process
		Long-term incentives setting process
	Effectiveness appraisal process	Specifying the intent of the effectiveness appraisal process
		Process descriptions of the effectiveness appraisal process
		Process schedule of the effectiveness appraisal process
		Review and improvement process of the effectiveness appraisal process

Table 6.6b The Last Four SC Requisite-Organisation Sub-Processes

Level Three	*Level Four*	*Level Five*
SC requisite-organisation process	Code-of-conduct process	Guideline for setting SC norms, principles and standards
		Guideline for setting SC appropriate and expected behaviours
		Guideline for setting SC inappropriate and unacceptable behaviours
		Security, discreetness and confidentiality process
	Personnel inflow/outflow management process	Process for calculation of the up-to-date target number of SC personnel required
		Period opening number of personnel calculation process
		Losses during current period capture and recording process
		New hires requirement = (target number of personnel required – opening number) + current period losses
	Training and assessment process	Process for determining relevant content focussed curricula for training
		Training teacher skill and competency standards setting and assessment process
		Developing an active learning bias process
		Process for managing multi-faceted learning (e.g. face-to-face lectures, short videos, written case studies, hands-on models, group learning sessions, on-line learning modules)
		Coaching and expert support availability process
		Process for managing ongoing training program of sufficient duration (not 'one and done')
		Formal pre- and post-training assessment process
		Sign-on and security clearances dependent on training completion and satisfactory assessment results

(*Continued*)

Table 6.6b (Continued)

Level Three	*Level Four*	*Level Five*
	Process for defining ICT support systems requirements	Clear understanding of how ICT can be used to facilitate use and enhance requisite-organisation results Functional ICT support system specification and approval process Business case for ICT support system development and approval process Approved ICT support system project implementation process ICT support system review and improvement process

Position descriptions are important because they define a number of important dimensions of a role, including its purpose, major responsibilities and performance expectations, reporting relationships, specific work requirements and selection criteria. The position description for a SC planner follows:

Position description

Supply Chain: Aluminium mining, smelting and processing

Position Title: Supply Chain Planner

Position Scope: All activities to do with the preparation, testing, presentation, approval and dissemination of the monthly end-to-end supply chain sales and operations plan.

Reports to: Supply Chain Planning Lead

Status: Full-time (38 hours/week)

Location: Geelong, Victoria, Australia

Purpose of the Position: This position exists to competently and reliably produce the supply chain's monthly sales and operations plan.

Responsibilities and Duties: The Supply Chain Planner is responsible for all of the activities required to produce a high quality and relevant monthly sales and operations plan. Such activities include demand planning, capacity planning, inventory planning, despatch and distribution planning, optimisation model building and testing, data collection, data cleansing and data verification, optimisation model running, analysis of results, sales and operations plan document production, plan presentation and dissemination of the approved plan. The Supply Chain Planner will also be responsible for a formal annual review of the planning process, including the design and implementation of process improvement actions.

Performance Expectations: The on-time delivery of an error free, optimised and verified monthly sales and operations plan. Demonstrable planning process improvement results from an annual planning process review. The Supply Chain Planner will be accountable for the delivery of these results.

Relationships: Key relationships include the Supply Chain Planning Lead, sales and marketing personnel, finance, logistics, procurement, operations and information and communication technology (ICT) personnel.

Social Skills: The Supply Chain Planner is expected to be mature and responsible, honest, respectful and discreet, friendly, approachable, cooperative, willing and helpful. Very importantly, the Supply Chain Planner needs to be an excellent communicator.

Authorities: The Supply Chain Planner has the authority to communicate with all those mentioned under relationships previously and to request all necessary data and assistance needed to fulfil the role. Such authority extends to the recommendation for approval of planning hardware and software tools/systems necessary to fulfil the role properly.

Selection Criteria: *Technical* criteria includes a minimum education requirement of either a BBus (SCM) or BSc (Maths) or BCompSc. Minimum work experience in related roles is ≥ 3 years. Accreditations from industry bodies such as ASCM, CIPS or CSCMP would also be viewed favourably. *Social* criteria include maturity, honesty, respectfulness, discreetness, friendliness, approachability, cooperativeness, helpfulness, willingness, acceptance of personal responsibility and superior communication skills.

Salary Band: AUD 60,000 to 75,000.

Employment Conditions: Superannuation scheme, annual leave, annual leave loading, paid sick leave for genuine illness, compassionate leave, maternity/paternity leave and long-service leave (as per the organisation's policy).

Signatures:

Supply Chain Planner: ____________________________

Supply Chain Planning Lead: ____________________________

Date: ____/___/_____

A third detailed process example will now be considered, being that of governance of an *effectiveness appraisal process*. This process is detailed in Figure 6.3; however, there are governance considerations to this process as indicated in Figure 6.4.

There are three personnel categories referred to in Figures 6.3 and 6.4: *unsuitable*, *achieving job goals* and *going above and beyond*. Too many leaders, unfortunately, do not deal appropriately with people in the *unsuitable* category. This is a bad outcome both for the individual involved and for the SC.

A key goal of the *effectiveness appraisal* processes, of course, is to assist as many people as possible to progress from *unsuitable*, if that is the case, to the *achieving job goals* category to the *going above and beyond* category. The latter is a win-win position for everyone involved.

The issue of managing all SC costs never really goes away; so it is with SC personnel costs. The key processes used to manage SC personnel costs are shown in Table 6.7.

Not every SC addresses the issue of lived principles, values and behaviours in a comprehensive and serious manner. There may be fancy MBA type documents in such organisations that *espouse* a range of values and behaviours, but espoused values and behaviours do not really change a culture unfortunately. To achieve real change and improvement to this SC outcome requires therefore a genuine and sustained effort. The SC processes available to achieve such change and improvement are shown in Table 6.8. Again, the processes listed in Table 6.8 may look simple and straightforward, but their execution is anything but. For this reason, it is recommended that SC leaders recruit specialist help to assist with the execution of these vital SC people sub-processes.

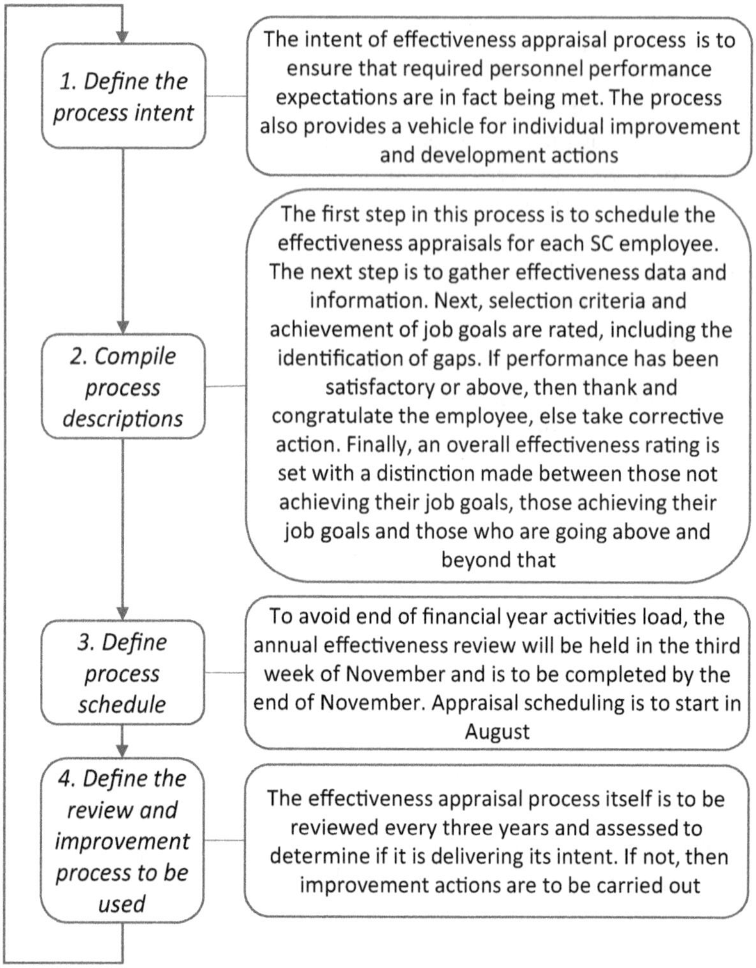

Figure 6.4 Governance of an Effectiveness Appraisal Process

Table 6.7 Processes for Managing SC Personnel Costs

Level Three	*Level Four*	*Level Five*
SC personnel costs	Tracking of SC personnel costs	Process for the determination and reporting of SC total personnel costs, costs per hour per classification, departmental costs, ordinary time costs, overtime costs, leave and entitlement costs, superannuation costs, recruitment costs, training costs, insurance costs, workers compensation costs, payroll tax costs, clothing/PPE costs, productivity levels
	Improvements to personnel productivity	Process for improving output per personnel hour from better tools and equipment, automation and continuous improvement to processes
	Personnel cost reductions	Process to assess and action levels of automation, process technology changes, better tools and equipment, job rationalisation and continuous improvement to personnel costs

Table 6.8 SC *People* Principles, Values and Behaviours Sub-Processes

Level Three	*Level Four*	*Level Five*
Principles, values and behaviours	Process for setting principles, values and behavioural standards and goals Process for building commitment to safety Self-respect awareness and development process Respect for others awareness and development process Commitment to delivering job goals building process Accepting personal responsibility awareness and development process Compliance with all job and organisational standards and principles assurance process Growing trust through exemplary behaviour process Ensuring fairness and equity process	Common process for expectations setting and deployment, training, coaching, monitoring, review, corrective action and improvement

Table 6.9 SC *People* Open Exploration and Sharing of SC Ideas, Collaborative Learning and Advancing SC Knowledge Sub-Processes

Level Three	*Level Four*
Open exploration and sharing of SC ideas	Ideas capture, appraisal and records system Successful ideas recognition and rewards process Process for the encouragement of team idea generation and idea assessment
Collaborative learning	Activating and sustaining SC improvement teams process SC communities of practice formation and management process Process for managing regular SC partner and customer performance improvement workshops
Advancing SC knowledge	Knowledge capture process for both internal and external SC knowledge Knowledge storage system specification and usage protocol Knowledge analysis, appraisal and interrogation process Knowledge dissemination process Knowledge system confidentiality, security and access protocols

The last three SC *people* level three sub-processes are those of *open exploration and sharing of SC ideas, collaborative learning* and *advancing SC knowledge*. Such processes oftentimes take a 'backseat' to the other SC *people* processes which is unfortunate because they hold considerable promise to the improvement of SC performance. Managed well, these processes as shown in Table 6.9, can and do deliver, at best, distinct competitive advantages and, at least, result in improved product and service offer, improved delivery reliability, reduced cycle times, strengthened SC resilience and flexibility and inventory and cost improvements. Real life examples include Arup, Cisco Systems, Coffey, DigitalOcean and HubSpot.

6.4 Dealing with supply chain *people* obstacles and barriers

There can be many, many obstacles and barriers to the competent management of SC *people* processes. Not ranked in any particular order, the obstacles and barriers typically encountered and how best to deal with them include:

Obstacle #1 – Shortage of suitably qualified, skilled and experienced people in the recruitment talent pool.

Suggested Response – Widen the search envelope by looking to other industries with similar skill sets, and/or build the required capability in-house if time permits.

Obstacle #2 – Poaching of staff by competitors.

Suggested Response – Provide a superior workplace culture to that of competitors, plus offer attractive performance incentives, plus make use of equitable employment contracts.

Obstacle #3 – Poor fits, that is wrong skills set for job or poor cultural fit (e.g. low performance ethic, low levels of personal responsibility, resistant to change, driven by dogma).

Suggested Response – For poor job fits, match skills set to another role and monitor performance. For poor cultural fits, conduct genuine attempt at cultural change; however, if that fails, then offer the individual(s) incentives to leave the SC.

Obstacle #4 – Bad, unacceptable, even toxic-like behaviour.

Suggested Response – Insist that the demonstrated unacceptable behaviours stop immediately. Set clear expectations and explain the consequences of non-compliance to expected and acceptable behaviours. Monitor and take action if the negative behaviours continue by exiting the individual(s) concerned from the SC.

Obstacle #5 – Absent and/or weak leadership (or worse, crippling leadership), plus lack of leadership 'buy-in' and support.

Suggested Response – Raise the issue with the leader concerned and explain calmly the ramifications of a continuation of the experienced leadership style. Respectfully request a change to a present, active and supportive style. If that doesn't happen, then request an interview with the leader's one-up. If that doesn't help, then think seriously about moving the leader to a different role.

Obstacle #6 – Silo mentality – a very strong functional bias along the SC.

Suggested Response – Seek the development of a shared SC vision, common purpose, common goals, open communications and functional collaboration via 'communities of practice'.

Obstacle #7 – Staff overload and burn out.

Suggested Response – Formal and professional project management of staff loading levels, including the identification of all personnel constraint resources (i.e. the ones who are or might become overloaded). Plus a formal health check process to monitor for signs of overload and stressed behaviour.

Obstacle #8 – Poor support systems. For example disconnected systems, poor interface design, inadequate functionality, slow response times, limited access equipment, outages and difficulties in making any sort of change. In short, everything that makes it frustrating for users trying to do their jobs.

Suggested Response – Apply all of the processes described in Chapter 4, SC *design* processes, specifically the design of SC ICT systems.

Barrier #1 – Equity targets. For example personnel selection based on diversity and equity targets.

Suggested Response – In the situation where all candidates are assessed as suitable then select to assure achievement of the equity targets. Where the capability gap for any given candidate(s) is too large (i.e. a candidate or candidates are well short of the skills, qualifications and experiences necessary to competently fulfil the role under consideration), then select from the remaining suitable candidates. If no candidates are deemed suitable, then relaunch the recruitment and selection processes.

Barrier #2 – Union domination of the workplace via legally enforced conditions.

Suggested Response – This is a very difficult situation to deal with. It is important in such circumstances, not to deal with union officials alone (i.e. SC employees need to be part of any discussions, consultations and negotiations as well as union officials). It is important also to try and reach a workable *accord* with the unions and employees. This may require extensive and tiring communications and negotiations; therefore a well-thought-out strategy, game plan with desired end points, all need to be established up-front. Discipline and perseverance are crucial to this process. If after three genuine attempts at improving the situation nothing has changed, then the organisation may need to consider exiting this particular SC.

Barrier #3 – Zero or insufficient funds exist to develop and competently execute all the necessary SC people processes.

Suggested Response – Never give up! Make (or remake) the case for SC people process excellence. Patiently explain the consequences of not doing these processes well. Provide evidence via examples to back up the case. Continuously request adequate funding until it is made available. If no funding is provided after numerous and documented attempts, then look seriously for another SC to work in.

6.5 Chapter 6 case study – achieving supply chain *people* excellence in practice

The case study for this chapter involves research conducted in New Zealand (NZ) on the factors blocking or hindering SCM and the quest for SCE. Also considered were the main enablers of the same (Luo et al., 2018).

The researchers decided to use a Delphi research method (expert panel type) in collecting data and information for their analysis. Two expert panels were formed – an '*academic*' group of 33 academics and a '*practitioners*' group of 107 practitioners, with all participants having knowledge and experience in operating NZ businesses. There were 59 participating companies across 17 NZ industries used in the study, with each of the companies operating within a wider SC.

Polling of the expert panels was conducted over three rounds. The first round identified a number of SC performance barriers and enablers. In all, 15 barriers and 17 enablers were identified. These groups were later reduced to five barriers and four enablers respectively. The second poll round established the importance of the nine factors as judged by the panels. The third poll confirmed the barriers and enablers and their rankings.

The *five barriers* identified, in order of importance (most important first), were:

i Lack of understanding of SC collaboration (also referred to by the researchers as integration). This was described as demonstrating a functional silo mentality, a lack of understanding about SC collaboration and a lack of trust between SC partners

ii Lack of internal buy-in to SC concepts. In turn driven by weak top management support, poor understanding of SCM generally and even misinterpretation of SCM concepts

iii A shortage of skilled SC staff and a low level of understanding of modern management methodologies

iv A culture that was resistant to change, influenced by previous negative experiences with SCM coupled with a weak focus on innovation

v Specific issues with the NZ business environment. This included exchange rate volatility, small domestic market size and geographic isolation, inefficient in-country infrastructure and government regulations

The four *enablers* identified, again in order of importance, were:

i SC collaboration. Specifically, the development of a lived SC *culture* – better end-to-end communications, building trust, an external focus and common SC goals
ii Staff education and training in SCM. This included all of the organisational levels that influence the adoption and management of SCM
iii End-to-end SC alignment. This included building a shared customer focus, a longer-term SC strategic focus, promoting the exercise of judgement within boundaries, active change management, benchmarking and performance measures
iv The application of advanced ICT. For example, to assist with internal to each company collaboration and SC-wide collaboration and the establishment of ICT standards so that different systems can 'talk' to one another

The researchers concluded by noting that people management issues are *the* most important in the achievement of SCE. This starts with top management support followed by the availability of competent SC practitioners. Next in line are the issues of overall SC culture and the development of a positive work ethic, acceptance of change and a spirit of innovation. This culture, in turn, enhances trust, communication and ultimately collaboration.

Lastly, the researchers made the point that, whilst the NZ SCs studied would benefit from additional use of ICT, this was not seen as a priority at the time of the study. It was suggested this additional ICT use may be more appropriate when the usage of SCM concepts matures. Additionally, it was considered that the use of modern and standardised SC performance measurement and continuous improvement processes would also be beneficial if applied.

The case study is very relevant to the content of this chapter as it reinforces many of the points of emphasis made throughout the chapter's text.

6.6 Review questions

i Why is it that the SC *people* processes can be considered as a *gestalt*?
ii Discuss possible reasons as to why some organisations take SC *people* process management very seriously whilst others do not
iii What is the difference between *espoused* and *lived* principles, values and behaviours? What impact does each type typically have on SC performance?
iv Discuss the *criteria elements* that can be used in the selection of suitable SC leadership candidates. From your real-world observations, are some of these selection criteria elements more important than others?
v Why is the existence of carefully crafted SC *position descriptions* so important to SC performance? What key factors must be included for a *position description* to be considered high quality?
vi What are the *effectiveness appraisal* outcome categories? How should SC people in each category be managed?
vii Discuss typical *obstacles and barriers* to implementation of sound SC *people* processes. What steps can be taken to successfully respond to such *obstacles and barriers*?

6.7 Assignment topics

(Each to be 3,000 words minimum or about three A4 pages single spaced. Tables and figures are not to be included in the word count.)

- i From your own research, choose two organisations that manage their SC *people* processes well and two that manage them not so well. Explain the differences in each organisation's approach and, importantly, explain the differences in overall SC performance for each case
- ii With the aid of diagrams and tables, define a process that SC leaders could use to effectively execute the '*leveraging the potential of all SC resources*' sub-process. In your description, include a listing of the inputs, the activities that must be undertaken and the process outputs
- iii From your research, choose *one* organisation that applies the *requisite-organisation* methodology well. What are the key features of your chosen organisation's approach and usage of the methodology? Has the adoption of the methodology helped or hindered the organisation's SC performance? Explain your answer
- iv You are a SC professional working in an organisation that is one of the partners in an electronics SC that spans three continents. The trouble is, your SC doesn't act in any way as a modern SC should, that is, there is no SC *culture* as the partners are more adversarial to one another, communications are minimal and information sharing is very guarded. Moreover, the level of understanding of SCM and SC process management is very low. There is almost a lethargy against trying anything new. Describe what actions you might take in your role to change this predicament, making sure you address each of the issues mentioned one at a time. Also indicate the sequence that should be followed for your recommended actions

6.8 Chapter 6 reference

Luo, W., Shi, Y. V., Venkatesh, V. (2018). Exploring the Factors of Achieving Supply Chain Excellence: A New Zealand Perspective. *Production Planning & Control*, 29(8), pp. 655–667.

7 Tackling supply chain *sustainability* head-on

7.1 What you will learn in this chapter

The chapter starts with an overview of sustainable development and a listing of and discussion on the United Nations' (UN) 17 Sustainable Development Goals (SDG).

Next is a detailed analysis of just what SC sustainability is, a listing of the main driving bodies and the main driver of sustainability, plus consideration of the typical barriers, risks and possible solutions to each.

Importantly, the risks of an organisation or a SC doing nothing about sustainability are highlighted.

On the basis that SCs really need to take SC sustainability seriously if they are to survive, a description of how SCs can best respond to the sustainability imperative is provided.

Answers to the question of how to conduct a sustainability cost/benefit analysis, followed by a case study of ten SC sustainability exemplar organisations is then presented.

The chapter concludes with review questions and assignment topics.

7.2 What specifically are the United Nations' sustainable development goals?

Firstly, it is important to define just what sustainable development means. In 1987, the World Commission on Environment and Development (WCED) proposed that it means '*a development that meets the needs of the present without compromising the ability of future generations to meet their own needs*'. Whilst this definition is rather broad and questions have been raised (Linton et al., 2007) concerning the exact nature of the needs of future generations and exactly what present actions and to what scale would compromise those future needs, the definition has nevertheless withstood the test of time.

At the UN sustainable development conference held in Brazil 2012, it was agreed to replace the UN's previous Millennium Development Goals (MDG) with a new set of global goals to be labelled the Sustainable Development Goals (SDG). The SDG were designed to address the environmental, political and economic challenges facing the world.

The UN 2015 SDG comprise 17 goals and 169 targets. The 17 goals are:

- i No poverty
- ii Zero hunger
- iii Good health and well-being

iv Quality education
v Gender equality
vi Clean water and sanitation
vii Affordable and clean energy
viii Decent work and economic growth
ix Industry, innovation and infrastructure
x Reduced inequalities
xi Sustainable cities and communities
xii Responsible consumption and production
xiii Climate action
xiv Life below water
xv Life on land
xvi Peace, justice and strong institutions
xvii Partnerships

At a high level, the UN SDG aim to guide global, regional and national efforts to reduce poverty, address climate change and build inclusive societies.

Criticisms of the UN SDG, whilst recognising the extensive consultation that went on leading up to their publication, is that they are more like a laundry list – too long and undifferentiated to one another. There is no guide to the relationships between the various goals, plus the goals are not all mutually exclusive. For example, 'gender equality' is a sub-set of 'reduced inequalities'. So why gender inequality alone was singled out over all of the other inequalities that exist in the world is not clear. Additionally, there is no prioritisation and no sequencing specified with the SDG, and there is no master plan, no roadmap or set of instructions to guide compliance or achievement. There is no coordination of the individual sustainability efforts that are being carried out globally.

It is considered that the goals need to be prioritised per region or area, that governments involved should use the prioritisations to drive policy, regulatory and tax reforms in order to provide the incentives for capital investment in initiatives and projects that will deliver the prioritised goals. All of this will be a major undertaking with many risks involved.

All of the SDG are relevant to sustainable supply chain management (SSCM) in some way. It is really up to individual SCs to prioritise the long list of goals and decide on the manageable set that it is able to pursue per time period. Attempting to work on the fulfilment of all 17 goals at once would not only be beyond the vast majority of SCs but it would also be fraught with risk of failure in taking on too much at once. This is not to say that pursuit of the goals is not an imperative, rather that professional planning (prioritisation, sequencing, resourcing, funding, constraint management and tight project management) is considered essential for real progress to be made.

7.3 Just what does supply chain sustainability mean? What are the key supply chain sustainability elements?

The word 'sustainability' means the ability to maintain some action or activity at a certain rate or at a certain level. This maintenance of action or activity can only be achieved, of course, if the *effects* of conducting the action or activity do not impede its continuation.

SC sustainability thus is the maintaining of SC operations in such a way that does not restrain or threaten its continuation. Any action taken by the SC or output from the SC (product or by-product) that causes internal or external damage beyond a sustainability limit will interfere with the SCs continuation and indeed may well lead to is discontinuation.

More specifically, SSCM can be defined as *the fulfilment of customer requirements via management of SC processes to effectively and efficiently manage the flow of materials, services, information, value-add and money in both a forward and reverse direction along the end-to-end SC whilst attaining target levels of performance on the three key goals of social, environmental and economic performance.*

The next logical question, given we understand what SSCM is, is why is it necessary? What are the drivers of SSCM? The main driving bodies of SSCM are illustrated in Figure 7.1. As can be seen, the *major driver* of SSCM thus is *meeting the sustainability expectations* of *customers, relevant governments* and *communities* (Seuring and Müller, 2008). Of course, other bodies, such as the UN, influence and facilitate each of the three main driver groups as well.

Importantly, and as also shown in Figure 7.1, the internal partners to any given SC are also drivers of SSCM as the delivery of sustainability initiatives can have a very positive effect on SC performance such as energy reductions, waste reductions, lead-time improvements, cost reductions from recycling and reuse, productivity improvements from the adoption of new and better technologies and reputation enhancement.

Given we have identified the major driver and main driving bodies of SSCM, a third logical question is, are there any barriers to SSCM – if there are tangible benefits to be obtained from SSCM as claimed, then why hasn't it been adopted universally? The answer is that there are a number of *barriers* involved with the adoption of SSCM that have contributed to a variable rate of take-up of the imperative. Such barriers, along with their associated complications and possible resolutions, are captured in Table 7.1.

In addition to the barriers to SSCM, as explained in Table 7.1, there are also a number of *risks* associated with the implementation of SSCM initiatives, and these are summarised, along with their respective complications and possible resolutions, in Table 7.2. There are, of course, substantial risks in *doing nothing* about SSCM, as discussed in Section 7.4.

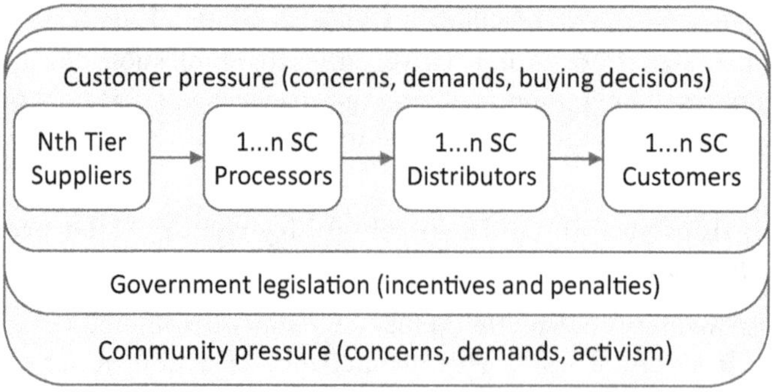

Figure 7.1 The Main Bodies Driving Sustainable Supply Chain Management

Table 7.1 Barriers to Sustainable Supply Chain Management

Barrier	*Complication*	*Resolution*
Not a conscious organisational priority	VUCA type business environment, low levels of SSCM ownership	Include as key part of SC strategy with specific *social, environmental* and *economic* goals. Embed in SC management system. Promote deliberate strategy of sustainable product development
Lack of knowledge and experience in actually practicing SSCM	Not a formal part of SC leader training and education	Active UN and local government sponsored SSCM training and education programs
Thought to be complex, difficult to achieve and a resource drain	Fear of the unknown, fear of failure	Promotion of best practice approaches, site visits to successful SSCs
SC levels of collaboration seen to be too low to achieve any real progress	Entrenched positions, 'not invented here' syndrome	Launch formal SSCM collaboration initiatives and projects
Seen as an additional cost only	Other cost reductions will need to be found to compensate	Identification of all possible SSCM cost reductions plus full SSCM life-cycle analysis
How to measure the *social* and *environmental* sustainability dimensions is unclear and/or applicable data is unavailable	Assessment of the social and environmental dimensions can be quite subjective, and expectations vary by stakeholder group	Identify the social and environmental sub-elements that matter most to a particular SC, e.g. unemployment rate, equity of workforce participation, workplace health, specific emissions, specific pollutants, waste management and power/fuel consumption, set targets and measure

Table 7.2 Risks Associated with Sustainable Supply Chain Management

Risk	*Complication*	*Resolution*
Up-front costs from adoption of SSCM initiatives drives down SC competitiveness	Competitor SCs exempt from SC sustainability requirements	Promote the social, environmental and economic benefits of the SSCM initiatives, budget for the up-front cost impact, ramp up customer communications
Implementation of SSCM initiatives over-run on cost and time	Supplier/service provider insolvencies, loss of key SSCM personnel	Apply 'Project Readiness and Assurance' process
SSCM initiatives generate unexpected side effects	No immediate knowledge or experience exist to deal with experienced side effects	Conduct formal risk analysis, including response to take if any side effect(s) is experienced

(*Continued*)

Table 7.2 (Continued)

Risk	*Complication*	*Resolution*
Expected deliverables of SSCM initiatives not achieved	Applied SSCM methodologies not proven solutions	Test all methodologies using a controlled experiment or pilot program before formal application
Sustainability ground rules and/or targets keep changing	'Flavour of the month' sustainability theories emerge	Focus on completion of valid current SSCM initiatives before moving onto new ones

7.4 What are the risks of doing nothing about supply chain sustainability?

Whilst the concept of SC sustainability has been around for many decades, it precipitated as a conscious organisational issue with the publication of John Elkington's book titled *Cannibals with Forks: The Triple Bottom Line of the 21st Century Business* (Elkington, 1997). Since that time, sustainability generally (including SC sustainability) has been receiving increasing attention in the business world, culminating in the UN SDG of 2015.

Importantly, Elkington has since emphasised that the triple bottom line concept wasn't generated to be used merely as an accounting tool and certainly not to be used as a corporate 'balancing act' where the triple bottom line elements are to be traded off against one another. A new process technology, for example, that reduces pollution, recycles wasted by-products, reduces costs and creates jobs is a win on all three sustainability dimensions. In other words, the three sustainability goals are not necessarily mutually exclusive. Such a conclusion has indeed been verified by research undertaken by Rao and Holt (2005), whereby the researchers identified the existence of significant positive relationships between actions taken to 'green' SCs and resultant SC competitiveness and economic performance.

Elkington was hopeful in introducing the concept that it would spark deeper thinking about the use of capital and the distribution of capital returns. Further, that the application of the triple bottom line in practice would bring about a system change to next generation business principles, including a focus on actively addressing the requirements of all relevant stakeholders.

It is argued that for any operating SC today, to countenance ignoring the triple bottom line concept in the manner of its original intent or ignoring the UN SDG is fraught with risk. Why might this be the case?

It is suggested that there are *four categories of risk* involved in answering this question:

i *Big picture risks* – These arise from further damage to social, environmental and economic conditions. Such damage includes further rainforest destruction, atmospheric damage such as the ozone layer depletion, plastic particle pollution of the world's oceans, soil contamination, permanent climate change, exploitation of labour, a growing inequitable distribution of wealth and corporate insolvencies

ii *SC internal risks* – These include the missing out on potential benefits that the progression of sustainability concepts and practices would deliver. Such benefits include cost reductions from energy usage efficiencies, waste reductions, recycling and reuse and avoidance of soil decontamination and waterway clean-up costs, lead-time improvements, productivity improvements, employee morale and motivation growth and lastly, increased sales. From this perspective, SSCM can be viewed as a SC excellence initiative

iii *Marketplace risks* – These include reputational damage, leading to a loss of market share, loss of key personnel and loss of investors

iv *Statutory risks* – These include a growing likelihood that communities and administrators will move against SCs and organisations that such groups consider are not doing enough to assure their sustainability. This could include class-action type litigation, fines, orders or even a withdrawal of an organisation's licence to operate. Additionally, recalcitrant SCs would miss out on any incentives offered by administrators for exemplary sustainability performance

It is for these reasons that SC sustainability is today a legitimate and important undertaking that simply must be taken seriously and actively responded to with carefully thought-out strategies that are implemented professionally using competent resources and technologies. SCs that ignore this sustainability imperative therefore will do so at their peril.

7.5 How can supply chains best respond to sustainability requirements?

Hong Kong based Cai and Choi (2020) conducted research on SSCM in the textile and apparel industry. They found an unbalanced focus on the UN SDG with 'Responsible Consumption and Production', 'Clean Water and Sanitation' and 'Climate Action' all receiving attention, whilst 'No Poverty', 'Reduced Inequalities', 'Life Below Water' and 'Life on Land' were receiving the least attention. The authors did, however, identify several positive SSCM steps being taken in the textile and apparel industry such as:

i *Sustainable design* – This is being progressed via careful choice of more environmentally friendly materials (such as organic cotton, lyocell and soybean protein fibre), designed in garment reusability/recyclability and the selection of environmentally superior process technologies

ii *Sustainable dyeing* – Suggested improvements include the use of spin-dyeing, screen printing and the use of natural pigments such as extracts from black tea, red beets and onion leaves

iii *Sustainable sourcing* – This includes an insistence that all suppliers conform to environmental and social standards such as selection of eco-friendly materials, proper wastewater treatment, no child or forced labour, normal working hours, safe work practices, safe working conditions and equitable compensation

iv *Sustainable production* – In particular, this applies to actively managing water usage (amount and treatment), energy usage (amount and emissions) and land contamination

v *Sustainable retailing and consumption* – Importantly, the authors found from their research that not all customers place high importance on sustainable consumption. Some customers are only prepared to pay up to 20% for eco-friendly products over non-eco-friendly ones. Rather, it was determined that customers, in some countries, are more favourable to the purchase of lower cost garments that have been repaired and/or redecorated. The authors conclude that increased emphasis should perhaps be placed on lifting public awareness of the importance of sustainable purchasing and likewise on the messaging strength of eco-labelling

vi *Reverse apparel SC* – Several barriers were identified to the real-life practice of reverse apparel SCs. Examples include a large fashion retailer and consumer culture whereby garment recycling is not commonly practiced, a lack of recyclability and reuse features built into material selection and garment design, limited wider promotion of the environmental attractiveness of recycling and reuse, limited life-cycle assessment of the end-to-end apparel chain and finally the absence of a determined push to convert the apparel industry into a truly circular SC

Taking this very valuable work of Cai and Choi (2020) and acknowledging the insight made by Kleindorfer et al. (2005) that circular SCs foster sustainability, it is possible to compile both a circular SC framework and a collaboration network of the key SC participants needed to give life to the framework. The framework is shown in Figure 7.2 and the collaboration network in Figure 7.3. Such concepts, if applied in reality, would substantially progress any SCs attempts at *social*, *environmental* and *economic* sustainability improvements.

A key underlying tenet of SCM is that end-to-end SC collaboration provides the opportunity for SC performance improvements that are far greater than the degree of improvements possible from each SC partner acting alone. Carballo-Penela et al. (2018) reinforce this point where they demonstrate for a four-stage seafood SC that overall SC carbon emission reductions can be up to *three times higher* when a collaborative approach is used and the SC partners combine to work on reducing the energy requirements of the top users rather than focussing on a set percentage reduction for each partner. In short, whilst SC collaboration is absolutely necessary for SSCM performance improvement, it nevertheless needs to be focused on the SC factors that will deliver the most gain.

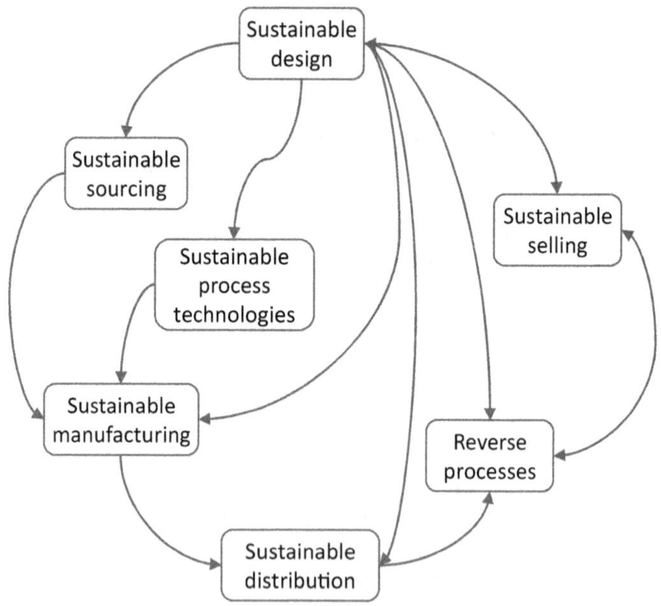

Figure 7.2 Sustainable Supply Chain Management from Circular Supply Chains

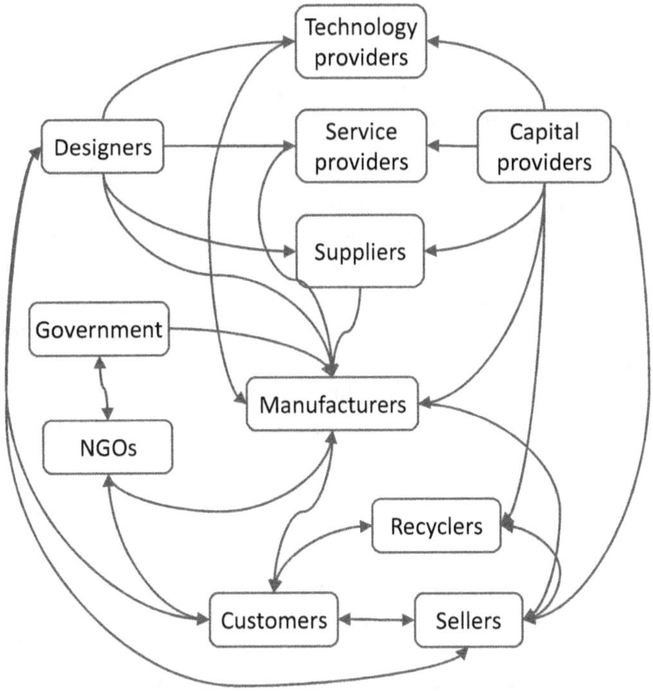

Figure 7.3 Sustainable Supply Chain Management Collaboration Network

An important SSCM implementation issue has been identified by Villena and Gioia (2020). The researchers studied nine top-tier and 22 lower-tier suppliers to three SCs, each headed by a multi-national corporation considered to be a 'sustainability leader'. The problem identified by the researchers was a diminishing of sustainability focus and attention as they moved from the sustainability leader back up the SC towards the nth tier supplier. The nth tier supplier problems identified included a lack of environmental management systems, a lack of procedures for handling red-flag social breaches, large numbers of temporary workers and rampant levels of overtime. The researchers emphasise the importance of:

- i *Consistent goals* requested on suppliers by various functions within the same organisation
- ii The need to improve *sustainability training* along the entire SC
- iii The need to provide each tier's procurement officers with *incentives* to achieve all three of the sustainability goals (and not just cost, quality and delivery alone)
- iv The *active engagement of suppliers* in auditing and continuous improvement initiatives

To conclude this section, and as correctly pointed out by Carter and Rogers (2008), all of the previously mentioned SSCM considerations need to start with formal SSCM goals, tactics and review mechanisms embedded in each SC's overall strategy. This needs to be complemented with a positive and supporting SC culture, emphasising the criticality of governance, ethical behaviour, transparency, genuine collaboration and effective planning for and response to all sustainability risks.

7.6 Can the cost/benefits of supply chain sustainability improvements be defined?

In order to answer the question as to whether expenditures on SC sustainability initiatives are of benefit and therefore can be justified, it is important to establish just how such potential benefits might be measured. Economic measures are reasonably standardised and have to comply with relevant country accounting standards. Unfortunately, there are no such standard measures for the social or environmental sustainability elements. It could be argued that the pinnacle economic measure of shareholder value-add (SVA) could just be copied for the other two elements – stakeholder *social* value-add and stakeholder *environmental* value-add. However, and unlike their economic counterpart, no commonly agreed derivation method or set of equations on how to calculate such terms, for either element, has been developed yet.

All is not lost, however, because it is still possible for organisations, within a given SC, to nominate a series of sub-measures that are relevant to their operations. Take the example of the aluminium processor AlyProc Pty Ltd. AlyProc's leaders wanted to get started with a proactive sustainability campaign but were surprised to find that no universally accepted overall measures existed for either the *social* element or the *environmental* element. The leaders decided to put the issue to their SC analysts and charged the analysts with the task of developing a recommended set of measures for *all three* sustainability elements. The results of the analysts' work can be seen in Tables 7.3, 7.4 and 7.5.

On observing the analysts' recommended set of measures, AlyProc's leaders were considerably impressed and remarked that such measures would be a solid starting point in pursuing sustainability improvements. The leaders requested that the analysts now start work on automating the data collection and regular reporting of actual results achieved on each of the measures. They also asked the analysts to consider how an *overall social sustainability index* and an *overall environmental sustainability index* might be developed.

It may well be the case that any given SC will want to understand the *very detailed* impact of its operations on the environment. In such cases, it is necessary to conduct a formal 'life-cycle assessment' of the SC's operations and the products or services it produces. As the term implies, such an assessment considers the cradle-to-cradle life cycle for each nominated product or service. This cycle is illustrated in Figure 7.4.

Life-cycle assessment has four main process steps:

i Setting the purpose, goals and scope of the study
ii Conducting an inventory analysis – This includes the identification of all of the environmental inputs and all of the environmental outputs involved
iii Impact assessment – The classification and evaluation of all environmental impacts
iv Interpretation – The verification of all conclusions arising from the study. This may include a third-party verification process

Whilst the previously mentioned four steps may appear straightforward, a formal life-cycle assessment process, as stipulated in ISO 14040 and ISO 14044, is a sizable undertaking if it

Table 7.3 AlyProc's Recommended *Social* Sustainability Measures

Social Issues	*Measures*	*Units*	*Last Assessment*	*Target*
Job availability	Regional Unemployment Level	%	6.30%	< 6.5%
Fair and equitable recruitment process	Approved process exists with formal scheduled reviews and updates	Yes/No	Yes	Yes
Fair and equitable selection process	Approved process exists with formal scheduled reviews and updates	Yes/No	Yes	Yes
Fair and equitable promotion process	Approved process exists with formal scheduled reviews and updates	Yes/No	Yes	Yes
Accredited training program	Accredited training program exists with formal scheduled reviews and updates	Yes/No	Yes	Yes
Accredited training assessment program	Accredited training assessment program exists with formal scheduled reviews and updates	Yes/No	Yes	Yes
Position descriptions	Approved position descriptions exist for each and every SC role, including formal scheduled reviews and updates	Yes/No	Yes	Yes
Intimidation	Approved policy exists with formal scheduled reviews and updates	Yes/No	Yes	Yes
	Number of genuine intimidation cases	No./employee/month	0.01	0
Harassment	Approved policy exists with formal scheduled reviews and updates	Yes/No	Yes	Yes
	Number of genuine harassment cases	No./employee/month	0.005	0
Safe work environment and safe work practices	Lost time injury frequency rate	Days lost/million hours worked	0	0
	Medical treatment injury frequency rate	No. of injuries/million hours worked	0.1	0
	Approved set of SC safety standards exists with formal scheduled reviews and updates	Yes/No	Yes	Yes
	Safety audit schedule compliance	% compliance	89%	100%
Employees in permanent positions	% of workforce genuine full-time	%	92%	95%
Working hours – eight hour days	Average hours worked/employee/week	Hours	52	< 48
Rates of pay	Percentile comparison to industry pay rates	Percentile	67th	80th

is to be done properly. For this reason, it is recommended that specialist help be engaged with any such formal life-cycle assessment attempts.

Returning to the main question of cost/benefit analysis, the added sustainability elements of *social* and *environmental* do make the appraisal of any SC projects or initiatives more complex than consideration of economic benefits alone.

Table 7.4 AlyProc's Recommended *Environmental* Sustainability Measures

Environmental Issues	*Measures*	*Units*	*Last Assessment*	*Target*
Atmospheric emissions:				
CO	Grams/kilogram of product output	Grams/kilogram	1.5	< 0.5
CO_2	*- ditto -*	*- ditto -*	125	< 50
NOx	*- ditto -*	*- ditto -*	57	< 25
SOx	*- ditto -*	*- ditto -*	22	< 26
Chlorofluorocarbons	*- ditto -*	*- ditto -*	0.5	< 2
Smoke	*- ditto -*	*- ditto -*	0.2	< 0.2
Dust	*- ditto -*	*- ditto -*	0.3	< 0.5
Soil and water contamination:				
Toxic waste – chromate	Grams/kilogram of product output	Grams/kilogram	1.2	< 0.2
Plastic	*- ditto -*	*- ditto -*	5	< 2
Chemical fertilizer	*- ditto -*	*- ditto -*	0.1	< 0.1
Pesticides	*- ditto -*	*- ditto -*	0.1	< 0.2
Herbicides	*- ditto -*	*- ditto -*	0.1	< 0.2
Fuel, oil, grease	*- ditto -*	*- ditto -*	1.2	< 0.5
Sludge	*- ditto -*	*- ditto -*	15	< 1.0
Other:				
Noise pollution	Noise level at property perimeter	dB	75	< 65
Lighting	Lighting interference beyond property perimeter	Instances	27	< 5
Road traffic congestion	Maximum road vehicles/ hour/route	Vehicles/hr	45	< 30

Table 7.5 AlyProc's Recommended *Economic* Sustainability Measures

Economic Issues	*Measures*	*Units*	*Last Assessment*	*Target*
For overall SC:				
SC costs	Total SC costs	$M	575	555
	Cost/unit of output	$/unit	27.3	25
SC days of inventory	Units of inventory/daily despatch rate	Days	32	21
Monetary value of all SC inventories	$ value of total SC inventories	$M	626	575
SC selling margin	(Selling price – unit cost)/selling price × 100	%	20%	30%
Net operating profit after tax	Revenue – full costs – tax	$M	262	270
% return on capital	Net operating profit after tax/capital employed	%	8.8%	10.2%
Shareholder value-add	Net operating profit after tax – $ value of cost of capital	$M	2.5	3.6
For SC projects:				
Net present value (NPV)*	Summation of future discounted cash flows minus project's capital cost	$,000	202	195
Payback period*	Time period taken for summed discounted cash flows to equal project cost	Years	2.1	3

* *For one selected project*

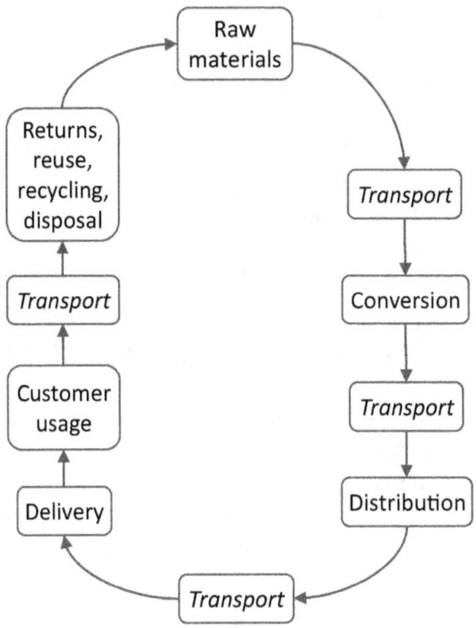

Figure 7.4 Physical Product Life-Cycle Assessment Scope

Any given SC project or initiative that delivers improvements to *social sustainability* and/or *environmental sustainability* and/or *economic sustainability* can have six main potential flow-on benefits:

i Compliance
ii Reputational
iii Customer relations
iv Community relations
v Employee relations
vi Economic

An *economic sustainability* project that has no economic benefit would obviously not be approved unless there were extenuating circumstances. However, it may well be the case that any given *social* or *environmental* sustainability project has no economic benefit but does have one or more of the other benefits. Such projects that do have flow-on compliance benefits would no doubt be approved; for with non-compliance, there is the very definite risk of fines or even losing the venture's licence to operate.

But what of a *social* or *environmental* focussed project or initiative that has no economic or compliance benefit? How can such proposals be assessed for approval? The answer is one of judgement – is the cost of undertaking this project or initiative (with no economic benefit) justifiable on the basis of expected long-run reputational, customer, community, employee relations and investor benefits? The answer depends on the scale of the expected benefits, the priority the organisation has placed on achieving them and the cost of the project. This really is decision making under uncertainty, and as such leaders would need to assess the probability of obtaining the identified benefit(s) and the long-run value of the benefit(s) to the organisation. They would also need to consider the consequences of *not* obtaining the benefit(s). Importantly, SC leaders also need to take into consideration any *performance gaps* identified via the sustainability measures, target setting, and actual performance monitoring process (example illustrated in Tables 7.3, 7.4 and 7.5).

It needs to be emphasised, of course, that many leading SCs today do take their *social* and *environmental* sustainability performance very seriously, as they do their *economic* performance. Examples include Sony, Philips, Cisco Systems, Merck KGaA, Colgate-Palmolive, Kering, Tatung, Johnson & Johnson and Schneider Electric.

The French luxury-goods corporation Kering (owner of Gucci and Balenciaga), for example, had by 2019 increased its traceability-to-origin of materials used in its products to 88% of its range. This is no small undertaking. Additionally, as early as 2015, Kering had developed and implemented an environmental profit and loss process (EP&L) that is used to access the environmental impact of each of its products.

The point is that there are by now many organisations active in the sustainability space and many innovative examples already exist for others to study and to emulate if they so decide.

Social or *environmental* focussed projects that *do* have an economic benefit can be assessed using the standard project evaluation approaches of net present value (NPV) and pay-back period.

Lastly and crucially, *sustainability* (including cost/benefit analysis) must be included as a key part of the management of each of the four key SC processes – SC *strategy* (Chapter 3), SC *design* (Chapter 4), SC *execution* (Chapter 5) and SC *people* (Chapter 6).

7.7 Chapter 7 case study – leading the way with supply chain sustainability

The case study for this chapter is based on research by Pagell and Wu (2009) of ten exemplar (leading) sustainable SCs operating in differing industries, of differing sizes and encompassing a range of manufacturing and service providers.

The fundamental research questions addressed were: *What is it that sustainable SC leaders are doing that is different? Are there patterns of behaviours demonstrated by the ten exemplars that could be used to develop a theoretical sustainable SC model?*

Clearly, the authors were trying to determine what made the ten exemplar SCs different from more 'traditional' SCs. In the process, they came up with a very elegant definition of what constitutes a sustainable SC: '*A truly sustainable SC could, customers willing, continue to do business forever*'. Whilst their research did not actually identify such a chain, they did nevertheless investigate ten SCs that were more sustainable than others in their respective industry.

Through a series of interviews with organisational leaders and personnel responsible for sustainability in each of the ten exemplars, site visits and third-party data collection, the researchers were able to piece together a fairly comprehensive appraisal of each of the ten organisations involved.

Using a combination of within-case and cross-case analysis, the researchers were able to identify SC sustainability enablers that had been previously identified in SCM literature and a series of enablers that had not. They were also able to identify patterns in the approaches used by the exemplars and derived an inventory of factors linked to SSCM.

The key findings of the research can be summarised as:

- i A culture of innovation and active committed leadership are each required for SC sustainability to be realised, but importantly if *both* attributes exist, then this positively amplifies the sustainability result actually achieved
- ii Sustainability performance is enhanced when *social*, *environmental* and *economic* goals are all *compatible* and are all *aligned*
- iii Sustainability needs to be *embedded* in the organisation's *culture* and thus become a *recognised and legitimate business activity* and one that is part of everyday *conversations*
- iv Remuneration systems that recognise and reward achieved performance on *all three* sustainability elements reinforce an ongoing focus on sustainability results
- v It is helpful if an organisation has an all-encompassing and keenly held *value*. This can be as simple as 'We ensure the well-being of all members of our SC' or 'We never outsource our pollution'.
- vi Supplier *continuity* is key to the realisation of a sustainable SC. Suppliers must not only survive and prosper but they also need to be a collaborative partner in the progression of sustainability initiatives. Engagement, training and active information sharing are all key to this factor
- vii Collaboration with *non-traditional* SC members such as NGOs, regulators, community members, even competitors, can enhance sustainability efforts
- viii SC sustainability can more successfully be built on the back of an already *well-performing* SC (as appraised by traditional measures) and less successfully built on the back of *poorly performing* SCs

In conclusion, the authors emphasise the importance of *active and committed leadership* to the SC sustainability imperative. Leaders' attention to non-economic performance was held as a high priority in the exemplar organisations studied, and non-economic goals were fully aligned with financial goals. Importantly, sustainability was found to be the way the exemplars undertook their business every day.

7.8 Review questions

- i Which of the 17 UN SDG are most relevant to operating SCs do you think?
- ii How could operating SCs best go about addressing the long list of SDG?
- iii Describe what you think the term SC sustainability really means? What are the implications of your definition for operating SCs?
- iv What are typical *barriers* to SSCM? How might such barriers be overcome?
- v What are typical *risks* to SSCM? How might such risks be overcome?
- vi What are the likely consequences if a SC does absolutely nothing about SC sustainability?
- vii How would you go about determining the cost/benefit analysis of any given SC sustainability proposal?
- viii What were some of the main factors found to positively influence SC sustainability outcomes in the chapter's case study?

7.9 Assignment topics

(Each to be 3,000 words minimum or about three A4 pages single spaced. Tables and figures are not to be included in the word count.)

i From your own research, identify an organisation that has attempted to deliver at least one of the UN SDG. Which SDG was chosen and why? How did the organisation's people approach the task and what results did they achieve? What are your key learnings from the case?

ii From your own research, describe the SC sustainability *strategy* adopted by at least one organisation or whole SC. What was the organisational driver for the strategy? Did the organisation (or whole SC) focus on *all three* sustainability elements? If yes, how did they ensure compatibility and alignment between each? If no, what did they focus on specifically? What results did they achieve?

iii From your own research, identify the *key measures* used by an organisation for each *social* sustainability, *environmental* sustainability and *economic* sustainability. How is the data collected on each measure, how is it analysed and how often is it reported? Does the organisation take any action as a result of the outcomes reported on each measure?

7.10 Chapter 7 references

Cai, T.-M., Choi, Y.-J. (2020). A United Nations Sustainable Development Goals Perspective for Sustainable Textile and Apparel Supply Chain Management. *Transportation Research Part E*, 141, pp. 1–21.

Carballo-Penela, A., Mateo-Mantecón, I., Alvarez, S., Castromán-Diz, J. (2018). The Role of Green Collaborative Strategies in Improving Environmental Sustainability in Supply Chains: Insights from a Case Study. *Business Strategy and the Environment*, 27, pp. 728–741.

Carter, C., Rogers, D. (2008). A Framework of Sustainable Supply Chain Management: Moving Towards a New Theory. *International Journal of Physical Distribution and Logistics Management*, 38(5), pp. 360–387.

Elkington, J. (1997). *Cannibals with Forks: Triple Bottom Line of 21st Century Business*. Oxford: Capstone Publishing Ltd.

Kleindorfer, P., Singhal, K., Van Wassenhove, L. (2005). Sustainable Operations Management. *Production and Operations Management*, 14(4), pp. 482–492.

Linton, J., Klassen, R., Jayaraman, V. (2007). Sustainable Supply Chains: An Introduction. *Journal of Operations Management*, 25, pp. 1075–1082.

Pagell, M., Wu, Z. (2009). Building a More Complete Theory of Sustainable Supply Chain Management Using Case Studies of 10 Exemplars. *Journal of Supply Chain Management*, 45(2), pp. 37–56.

Rao, P., Holt, D. (2005). Do Green Supply Chains Lead to Competitiveness and Economic Performance? *International Journal of Operations and Production Management*, 25(9), pp. 898–916.

Seuring, S., Müller, M. (2008). From a Literature Review to a Conceptual Framework for Sustainable Supply Chain Management. *Journal of Cleaner Production*, 16, pp. 1699–1710.

Villena, V., Gioia, D. (2020). A More Sustainable Supply Chain. *Harvard Business Review*, March–April, pp. 84–93.

WCED (World Commission on Environment and Development). (1987). *Our Common Future*. Oxford: Oxford University Press.

8 Preconditions for supply chain excellence success

8.1 What you will learn in this chapter

The first seven chapters of this book have presented a *roadmap* on how to achieve SC excellence through SC process excellence and how to do that *sustainably*.

A *roadmap*, however, is of little use if the state's borders are closed due to a pandemic or there is no suitable transport vehicle available or no fuel to put in the vehicle or no licenced driver to drive it!

This chapter, therefore, presents a description of the essential preconditions that really need to be in place if efforts to achieve SC excellence are to succeed.

To begin, a precondition summary is presented and importantly the interrelationships that exist between the desired outcome of SC excellence and the precondition clusters necessary to achieve it are illustrated.

Next, important *socio-preconditions* and *technical-preconditions* are outlined, followed by a description of each and the *possible solutions* that can be applied if any one of the preconditions is absent.

The interrelationships between the preconditions and SC sustainability are then described, including a line-of-sight diagram showing the system dynamics involved.

An end of chapter case study is presented, describing research undertaken on 122 Hong Kong based firms participating in recognised SCs. The case study provides good support for the material presented in this chapter.

The chapter ends with a set of review questions, written assignment topics and the chapter's references.

8.2 Preconditions for supply chain success and survival

There are many preconditions for the achievement of genuine SC excellence, leading to long-run SC success. Whilst the preconditions listed in this chapter are described in the positive, the absence of any one of the preconditions can be taken as a barrier to realising superior SC performance.

The preconditions can of course be summarised, and such a summary would read as:

> *The presence of a good leader, with a good team, good resources, good facilities and infrastructure, advanced technologies, attractive products and services, keen customers with good purchasing power, a 'blue-ocean' type competitive environment and adequate time to complete tasks and projects. Coupled with a strong and determined focus, persistence, a caring attitude and an ethos of innovation, sustainability and continuous improvement, then the ingredients of SC success exist.*

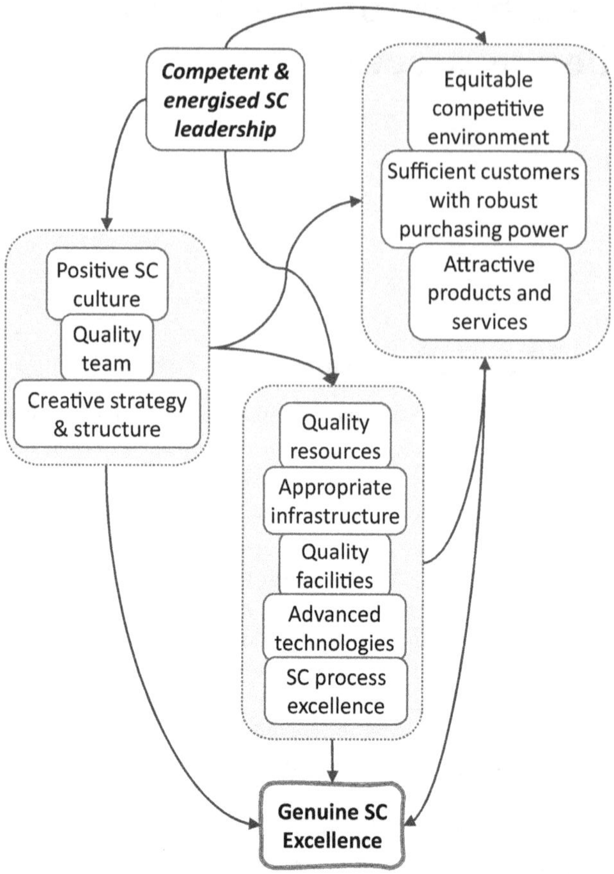

Figure 8.1 Essential SC Excellence Preconditions Summary

The preconditions summary can be visualised as shown in Figure 8.1.

Such a preconditions summary may seem like a pipe dream to many; however, with diligence and determination, such an SC environment can be created via careful attention to the individual preconditions listed in the chapter and through the implementation of relevant actions as suggested by the *solutions* proposed herein.

The preconditions can be grouped broadly into two groups – a *socio* group and a *technical* group as shown in Table 8.1.

Each group is now considered along with suggested solutions if the respective precondition is absent.

Socio-preconditions and solutions if absent:

i ***Strong, determined leadership with a genuine sustainability ethos*** – This is typified by leaders who are aware, grown-up and mature, stable, competent, experienced, present and active, visionary and determined, persistent, trustworthy, inclusive, friendly, supportive and collaborative

Table 8.1 Socio-Technical Preconditions for SC Excellence

Socio-Preconditions	*Technical-Preconditions*
Strong determined leadership with a genuine sustainability ethos	An approved set of aligned strategies, goals, targets and action plans
Two-way vertical free-flow of information	Attractive value propositions
Supportive and effective behaviours	Keen customers with real purchasing power
Team member competence, commitment and attitude	Equitable competitive environment
Alignment of organisational functions	Free-flowing channels to market
High levels of employee morale and motivation	Fulfilment of all customer requirements
Power and passion balance	Availability of quality personnel, raw materials, services, utilities and funding
Regular and quality communications	Proven technical solutions (for processes, equipment, data management & methodologies)
An active accountabilities hierarchy	Competent pre-launch planning for all initiatives & projects
Active effectiveness appraisals	High quality of resources, tools, metrics and analytic methodologies
	Active management of personnel working loads
	On-time deliveries of supplies and work tasks
	Preparedness for SC shocks
	High quality operating facilities
	Optimal location of operating facilities
	Sensible use of technology (e.g. Cloud computing, big data & data analytics, networks, process integration, mobility devices, fully digital SC, information transparency and traceability)

Possible solutions to the absence of strong, determined leadership include the undertaking of retraining, effectiveness appraisals and reviews, including the setting of clear leader expectations. Additional SC leadership assurance initiatives are discussed in Chapter 6. If such improvement actions fail to deliver the required behaviours after two genuine attempts, then leadership replacement will be necessary

ii ***The free-flow of SC information*** – This includes the free-flow of vision, strategies, policies, goals, expectations, tactics and initiatives down through the SC organisational layers and the free-flow of data and information concerning performance results, event descriptions, comparative performance levels, ideas, suggestions and exceptional individual performance back up through the SC organisational layers. Unfortunately, many organisations suffer from middle managers' 'suds-clog'. That is, the middle layers of management act as filters and blockers to the ready flow of data and information in order to protect their own interests. A good example of this was the experience of POSCO (a large South Korean steel producer) and its early attempts at the introduction of Six Sigma improvement methodology to its operating plants in 2002. So much middle management passive resistance was confronted that the business leaders found it necessary to have the Six Sigma teams report directly to them. Six Sigma turned out to be a major winner for POSCO, so much so that it now markets its own brand of 'Quick Six Sigma Plus' (QSS+) and helps its affiliated organisations implement it

Possible solutions to the absence of free-flow of data and information include clear descriptions of the free-flow expectations in each role's position description, the compilation of a professional responsible, accountable, consulted, informed (RACI) matrix where the positions that are *responsible* and *accountable* and those that are to be *consulted* and those *informed* are clearly spelt out. The mechanisation of data and information flow using advanced technologies, such as enterprise resource planning II (ERPII), in order to bring visibility and transparency of data and information to all those along the SC who need it to do their jobs properly. If such attempts fail to solve the problem, then delayering of the organisation may well be necessary

iii ***Supportive and effective behaviours*** – Organisational culture can either be positive or negative along a cultural rating scale. A positive culture engenders a sense of belonging for people, where feeling valued and respected flows from value-added achievements. A negative or 'toxic' culture that is resplendent with a long list of dysfunctional behaviours is a recipe for SC failure. Such dysfunctional behaviours include resistance to change, malicious compliance, malevolence, hostility and even straight-out corruption

Possible solutions to the absence of a positive culture include clearly describing, documenting and communicating the culture that is desired, consistently applying the desired culture by reinforcing expected behaviours and sanctioning unacceptable behaviours, regular ongoing communications including discussions of known culture performance gaps, influencing of leaders and finally a recognition that negative cultures do not usually stem from 'bad people' but rather are usually allowed to develop by 'bad leaders'

iv ***Team member competence, commitment and attitude*** – Teams perform well when the individual team members are very clear on the team's deliverables, know what they are doing, are energised and committed to doing it, and do it in a cooperative and collaborative way. If such a team culture is absent, the team will struggle to deliver their expected outcomes

Possible solutions to such an absence include taking the time to produce a tested and approved team task brief (task purpose, deliverables, scope, timeline, resources, team members, team leader, interrelationships, team protocols and expected behaviours), very careful selection of team members, team member training (or even retraining if necessary), attractive incentives to deliver a quality output on-time, clear team member re-entry paths (if working on a project for example) and clear expectations around levels of cooperation and collaboration

v ***Functional alignment, common purpose, shared values, open cooperation and transparency*** – Organisational functions are usually a considerable dilemma for SC leaders. On one hand, the functions are the repositories of essential functional knowledge (design, engineering, marketing, manufacturing, finance, ICT, SCM) whilst on the other hand they can and often do become organisational silos, developing very definite silo mentalities such as insular withdrawal, retention of data and information, separate goals and objectives, competitive rivalry and an uncooperative elite type attitude

Possible solutions to such a silo mentality are active and strong leadership from the position above the silos, the setting of common goals, the insistence that data and information be freely shared, the identification and living of shared values and mandatory participation in collaborative initiatives (e.g. communities of practice) and projects

vi ***High levels of employee morale and motivation*** – This includes a proactive engaging attitude, acceptance of personal responsibilities for the delivery of job goals, acceptance of role accountabilities, acceptance of necessary change and high levels of skills and competencies

Possible solutions to an absence of such attributes include building trust through honesty, consistency, respect, openness, inclusiveness and fairness; the provision of desired and worthwhile incentives, a safe, clean and attractive working environment, appropriate training for knowledge building, proper tools and equipment, space to perform, recognition for all positive achievements

vii ***Power and passion balance*** – SC leaders are not exempt from the dangers of the excessive use of power or of becoming passion compelled. Passion is needed of course else nothing will happen, but it needs to be balanced with level-headedness and rational thinking. Power is also important as authority is needed for statutory reasons, for access to funding, project and expense approvals and the organisation of work teams and positions. Excessive use of power, however, usually ends with people behaving badly, displaying, for example, aggression and/or recklessness

Possible solutions to excesses of power and/or passion include adopting a genuine attitude of humbleness, patience and personal discipline, an open dialogue to constructively discuss the issue, recognition of each individual's ability to balance passion with self-control and open and honest feedback as soon as any excesses are observed or detected

viii ***Regular and quality communications*** – This issue comes up almost always on any employee survey looking into what is required to improve workplace relations. In part, that may be because it is just so obvious to most people that communication is essential to any relationship. However, there may well be a darker side to the fact that it is raised so often. That is, there is every possibility the reason it is raised so often is because people consider that the level and quality of communications they are receiving is simply inferior

Possible solutions to communication deficiencies therefore include the collaborative development and adoption of a SC circular communications standard. Such a standard would need to identify the objectives and goals to be achieved – What specific information is to be communicated? To what level of granularity? To what length (e.g. factsheet, one-page quality story, presentation slides, brochure, 20 minute podcast, product video, one hour webcast, progress report, policy statement, detailed SC standard, monthly report, comprehensive report)? Who specifically has the responsibility for crafting the communication? Who specifically has the responsibility for deploying it? To whom or what specific groups is it to be communicated to? Is the communication schedule to be at regular specified time periods or is it to be milestone driven? What is the communication medium (e.g. face-to-face one to one, face-to-face one to many, face-to-face meeting or conference, video conference, tele-conference, webcast, podcast, commercial media, email, SMS message, webpage document, hard copy)? Is the communication to be verbal only or written only or both? What mechanisms are to be used to encourage and capture all feedback? What are the processes to be used that will ensure follow-up of such feedback? Finally, it is crucial to have active communication metrics system in place that is used as part of a continuous improvement loop

As can appreciated from this description, quality communications are a non-trivial undertaking and so much hinges on doing communications well. It is therefore not surprising perhaps that it receives so much attention in customer, supplier and employee surveys and in employee effectiveness appraisals

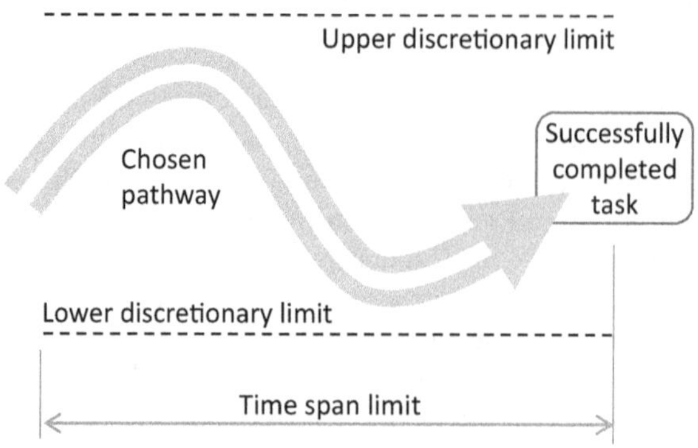

Figure 8.2 Successful Task Completion – Exercising Judgement Within Boundaries

ix ***An active accountabilities hierarchy*** – This is a key feature of the 'requisite-organisation' approach to people management, and it is essentially a framework that employees use to exercise judgement within predefined boundaries as illustrated in Figure 8.2. This approach is key to personnel performance management as attempts at empowerment have oftentimes led to anarchy and self-directed work teams to adhocracy where the minimum amount of work is carried out and only when absolutely necessary. An accountabilities hierarchy on the other hand does give employees space to act but still holds them accountable for observance of all relevant standards and delivery of their job goals and successful task completion

Possible solutions to the absence of an accountability hierarchy include the adoption of the 'requisite-organisation' approach to people management. As this is no small undertaking, it is recommended that reputable and reliable specialist help be engaged to assist in the design and implementation of this transition

x ***An active effectiveness appraisal process*** – Again, this is normally a key part of the 'requisite-organisation' approach. The purpose is to provide an accurate and relevant assessment of how each individual is performing in their position. Are they achieving their jobs goals or not? Are they going above and beyond? Important dimensions are the *quantity* of the work completed, the *quality* of the work completed, and the *time* taken for the *resources* used – the so-called task assignment QQT/R model. Managers can use this model to set targets for each of the QQT/R factors for each task performed and can also use the model to assess actual performance on those same tasks. Some people add CP (for context and purpose) to the beginning of the term and end up with the long mnemonic CPQQRT

Possible solutions to the absence of an effectiveness appraisal process are the same as for the previous point as effectiveness appraisal is normally part of the 'requisite-organisation' methodology

Technical-preconditions and solutions if absent:

i ***An approved set of aligned strategies, goals, targets and action plans*** – The elements of SC *strategy* along with specific examples are covered in Chapter 3 of this book

Possible solutions to the absence of a formal SC strategy or a deficient SC strategy include the activation of all of the SC *strategy* sub-processes outlined in Chapter 3

ii ***Attractive value propositions*** – Customer buying decisions are centred around the value offer made to them by the SC. If customers perceive the value proposition to be inferior, then they will not buy from the respective SC, so it will struggle to survive. The relationship between SC excellence and attractive value propositions is quite symbiotic in nature. That is, SC genuine excellence assures attractive value propositions and attractive value propositions provide the funds (via customer purchases) to build and maintain SC excellence

Possible solutions to the absence of attractive value propositions include establishing just what it is that the SC's customers value and then taking the necessary steps to develop and offer such product and/or service attributes to them. This really needs to be a circular process because not only do customer value expectations vary over time so too do the new offers made to customers by the SC's competitors

iii ***Keen customers with real purchasing power*** – This is related to the previous point – customers will be keen if they feel the SC is offering them value, but they will be even keener if they can afford to pay for the product or service they need or desire

Possible solutions to the absence of keen customers with real purchasing power include investing sufficient funds to allow the detailed analysis of customers' needs, wants and desires and then carrying out the product and/or service research necessary to fulfil the identified customer requirements and the capability of offering it at a price customers can afford. This invariably means segmentation of the market offer as some customers are willing and able to pay more than other customers

iv ***Equitable competitive environment*** – Prima facie, it may appear that this is outside the circle of influence of SC leaders and the SC overall. However, SC leaders can still take actions to influence the competitive environment, especially if the present competitive environment is characterised by hostile and aggressive competition

Possible solutions to the absence of an equitable competitive environment include a carefully planned, well-executed strategy aimed at moving further along the *differentiation ←→ commodity* continuum than is presently the case. This may well require some very big decisions, particularly if the bulk of presently offered products and/or services are at the commodity end of the continuum. Cirque du Soleil is a perfect example of this, whereby it left behind the tired old circus model to develop a new, unique and differentiated form of big-top entertainment. Another possible solution is to grow brand image and reputation by offering reliable fit for purpose products and delivering them faultlessly whilst offering customers something that is unique and appeals to their beliefs. An example of this would be a commitment to a promotional statement such as: 'We stand for sustainability'. This would only work of course so long as the SC actually does stand for sustainability. A third solution would be the development of products and services that are patentable. Pharmaceutical companies are good examples of this practice. Patents have but a finite life of course, so the development of new patentable products is an ongoing requirement. Lastly, a SC may decide to exit its current industry and migrate to a completely new one. General Electric (GE), for example, announced on 21 September 2020 that it is to progressively exit the coal industry. This is a major decision that will have many flow-on effects

v ***Free-flowing channels to market*** – Clear access to markets is a key requirement for a successful SC. One of the more problematic circumstances a SC can find itself in is one where the channels to market are controlled by external interests. Such a situation can lead to increased costs, preference given to transfer of competitors products or even an inability to ship orders to customers. A classic case of this arose in the Pilbara region of Australia in 2009 when start-up iron ore producer Fortescue Metals wanted access to rail lines owned by BHP and Rio Tinto in order to ship its mined ore to Port Hedland on the West Australian coast, rather than building its own line. BHP and Rio Tinto resisted this of course because it would cause interference to the shipment of their ore. Fortescue was eventually granted the right to appeal to the Australian Competition and Consumer Commission if an agreement over access terms and conditions could not be reached. In 2010, however, the Australian Competition Tribunal ruled that Fortescue could access only two of the three lines in the region. In the end, Fortescue spent some A$2.5 billion and built its own line, finishing construction in 2012. Such is the importance of channels to market!

Possible solutions to the absence of clear channels to market include entering into agreements or alliances with those organisations controlling the channels or purchasing them or building new separate channels such as Fortescue Metals did

vi ***Fulfilment of all customer requirements*** – Major customer requirements include perceived value arising from attractive and fit for purpose products and services that can be purchased for competitive prices and delivered in-full, on-time, with short order lead times and error free. Other requirements include on-line pre-sales services such as accurate and available product and service information, ease of ordering, order tracking, home delivery, delivery confirmation and product in-service reviews

Possible solutions to the sub-standard fulfilment of customer requirements include the actioning of all the sub-processes outlined in the 'design for service' section of Chapter 4 and the 'customer service' and 'order fulfilment' sections of Chapter 5

vii ***Availability of quality personnel, raw materials, services, utilities and funding*** – Such requirements are essential to any SC excellence quest, and whilst they are straightforward to write about, they are usually very difficult to achieve in practice, primarily because there is so much competition for them

Possible solutions to the absence of such essential factors include the straight-out purchase of them or procurement through alliances or very close and strategic relationships with the suppliers of such essentials. In regard to quality personnel in particular, it is recommended that all of the SC *people* processes, explained in Chapter 6, be adopted and put into practice

viii ***Proven technical solutions for all processes, equipment, data management and SC operating methodologies intended for use*** – Fiat Chrysler installed an unproven gear shift design on a number of its vehicles between 2012 and 2015 (Lawrence, 2019). The design problem has been linked to 266 rollaway incidents, 68 of which resulted in injuries. The company recalled the vehicles in question, but the new 'fix' was also found to be problematic. Such situations are expensive to rectify and are made worse if damage and injuries are involved. Even more serious perhaps is the potential for reputational and/or career damage

Possible solutions to the absence of proven solutions include the adoption of a simple golden rule – '*If it is not proven, we don't use it*'.

ix ***Competent pre-launch planning for all initiatives and projects*** – So many good and needed projects have failed because of inadequate pre-launch planning. Failure rates on new projects due to shortfalls to on-time delivery, promised functionality or cost can be as high as 43%. One in six ICT projects, for example, fall into the 'black swan' category (Bloch et al., 2012)

Possible solutions to the absence of adequate pre-launch planning include the rigorous use of the *project management and assurance process* as explained in detail in Chapter 5

x ***High quality resources, tools, metrics and analytical methodologies*** – This requirement can best be summed up by the old saying: 'Garbage in equals garbage out'. That is, the best conversion processes in the world will still produce inferior outputs if the inputs are inferior. The same applies to procedures, metrics and methodologies. Improvement methodologies such as Lean, Six Sigma, ECR, CPFR, TQM and the ISO 9000 family of quality standards are all by now well-established and proven. So long as they are applied properly and maintained, then organisations using them can have high levels of confidence about their effectiveness and value-add

Possible solutions to the absence of quality inputs and quality methodologies include the establishment of quality management processes for all inputs. This includes supplier and service provider accreditations, audits of suppliers' and service providers' quality assurance systems and procedures, the insistence that suppliers and service providers provide product and service test results and, as a last resort, the inspection and testing of deliveries from suppliers and service providers. For methodologies, it is strongly recommended that only proven solutions be chosen and when attempted for the first time, that specialist help be engaged to ensure the proper use of the methodology

xi ***Active management of personnel working loads*** – A trap that many project teams fall into is the failure to recognise that individual team members can and often do become constraint resources for the team. A constraint resource is a bottleneck, and when a bottleneck becomes overloaded, it results in late deliveries and stress. People bottlenecks occur when a team member with a unique skill set or who holds a gatekeeper role is called upon by several project components at the same time. Non-bottleneck human resources can also be overloaded if their work schedule is not carefully managed

Possible solutions to the absence of active workload management include the adoption of reliable project management methods such as the theory of constraints based 'Critical Chain' approach (Goldratt, 1997). A project management and assurance process example is presented in Chapter 5

xii ***Assurance of deliveries*** – This includes supplies, services and work tasks. Late deliveries of either physical goods or non-physical services can be very disruptive to any work team, business unit or SC. Steps need to be taken, therefore, to ensure delivery assurance of such inputs

Possible solutions to the absence of delivery assurance include (a) the setting of positive and negative delivery performance consequences, (b) a stipulation that safety stock mechanisms be employed in order to buffer supply side variations and disruptions, (c) assisting suppliers and service providers in setting up and using a professional project management process and (d) direct involvement with suppliers and service providers in the development of and auditing of the order fulfilment processes they use

xiii ***Preparedness for SC shocks*** – Low probability events that have a high consequence can have disastrous impacts on supply chains as witnessed by the 2011 Japan earthquake and tsunami and the 2020 COVID-19 pandemic. Because of their low frequency of occurrence and difficulties in forecasting their level of intensity, the level of preparedness for such SC shocks is not always as good as it needs to be

Possible solutions to the absence of adequate preparedness include the utilisation of the SC *resilience builder* approach as explained in Chapter 2 and, specifically, as shown in Figure 2.8

xiv ***High quality operating facilities*** – The word 'quality' in this precondition means operating facilities that are safe to operate, that operate reliably, that produce defect free products and that are sustainable. SC excellence is not possible to achieve if a venture's operating facilities keep breaking down and/or keep producing defective products or services and/or are unsustainable from an environmental, social or economic perspective

Possible solutions to the absence of quality operating facilities include the restoration of existing low quality operating facilities or, if that is not possible, their replacement. New facilities need to be verified as proven solutions from demonstration site visits and audits, potential suppliers need to be reputable and accredited and all facility performance requirements need to be carefully defined and documented. Facility construction and installation needs to be managed via a competent project readiness and assurance process, operator training needs to be formal and individual certification awarded only if post-training assessment is satisfactory. Commissioning practices need to be best practice and post-commissioning testing needs to be carried out before facility handover to operations

xv ***Optimal location of operational facilities*** – This precondition really highlights the crucial nature of SC design. Once operating facilities are 'in-the-ground' it can be very costly and disruptive to change them. Locating a major distribution centre hundreds of kilometres away from retailers and consumers just because the land in the centre's chosen location is cheap is one of many examples of inadequate planning and design around location decisions. Land price is of course important but is only one of many factors that must be assessed in such location design decisions

Possible solutions to the absence of facility location optimality include the introduction of a formal SC *design* sign-off in all project or initiative business cases. This means that those submitting the business case must demonstrate the results of extensive facility location optimisation testing and the details of the decision analysis used to choose between the different options available. After analysis, the recommended option (preferred design) needs to be agreed, approved and signed-off by those approving the project or initiative. This process needs to apply to all greenfield site decisions and to all facility relocation decisions

xvi ***Sensible use of technology*** – This includes such technologies as Cloud computing, big data management and data analytics, high bandwidth networks, process integration, mobility devices, fully digital SCs, information transparency and traceability. This topic possibly represents one of the biggest opportunities for SCs to differentiate themselves in the hearts and minds of their customers. Such systems and methods, for example, allow for the effective functioning of an omni-channel marketing strategy; that is, the marketing, selling and serving of customers in a seamless way across all of the SCs channels and methods of interacting with customers. So, no matter if a customer orders on-line or by telephone or in-store, across different channels, the experience will be the same. So too will be all of the post-order interactions such as order status reporting, deliver enquiries, deliver confirmation and in-service ratings. There are many other benefits available from the sensible adoption of ICT and big data technologies, including process improvements, supplier 'health' assessment and full product or service traceability

Possible solutions to the absence of technology adoption include building the case for its adoption and convincing SC leaders to participate in the assessment of the potential benefits and value-add from such adoption. This may require site visits to organisations (perhaps in different industries) that have enthusiastically embraced such technology and where it works well. The proper management of big data, for example, can be a real selling point because proper storage and analysis practices for big data can help reduce data storage costs. Additionally, the application of analytics to big data can lead to the identification of substantial improvement opportunities, as described in the book *Supply Chain Analytics* (Robertson, 2021). Lastly, in addition to potential capital and cost reductions, the adoption of Cloud computing offers, at last, the real opportunity of a SC-wide ERPII system that could underpin SC collaboration unhindered by data and information unavailability (Koh et al., 2011)

8.3 Relationship between supply chain excellence preconditions and supply chain sustainability

As described in Chapter 7's case study, one of the conclusions reached by Pagell and Wu (2009) in their research of ten exemplar SCs was that SC sustainability can be more successfully built on the back of an already *well-performing* SC (as appraised by traditional measures) and less successfully built on the back of *poorly performing* SCs.

This conclusion of the researchers makes logical sense. That is, personnel working in a poorly performing SC would no doubt be very busy trying to fix their poor performance or alternatively are simply incapable of fixing it. Either way, their capacity and/or capability to take on all three of the sustainability goals and meet them, in addition to meeting their other SC goals, is highly unlikely.

By comparison, SCs that are performing well would undoubtedly to be staffed by competent and disciplined professionals using proven methodologies and capable facilities. Such operatives would be much better placed to respond to the task of achieving set targets on each of the sustainability dimensions – social, environmental and economic.

As illustrated in Figure 8.1, the achievement of SC excellence is dependent on a number of preconditions being in place and working effectively. Taking those same preconditions, Figure 8.3 provides a line-of-sight between the precondition clusters and the three sustainability goals. As can be seen, they are highly interrelated, and one of the implications of this level of interrelationship is that if one of the clusters of preconditions is taken away or is not performing well, then the whole system suffers.

In complex systems, such as the one shown in Figure 8.3, it is often equally important to understand all of the systems relationships that exist as it is to understand the individual system components. In such systems, the behaviour of the whole cannot necessarily be ascertained by studying the behaviour of the individual parts alone. For example, competent and energised SC leadership is critical to the obtainment of SC excellence as it is to achieving SC sustainability goals. SC leadership, however, does not exist in a vacuum, and it cannot *alone* ensure delivery of the beforementioned desired SC outcomes. All four of the precondition clusters (including competent and energised SC leadership) need to be in place and working effectively for the goals sought to be realised.

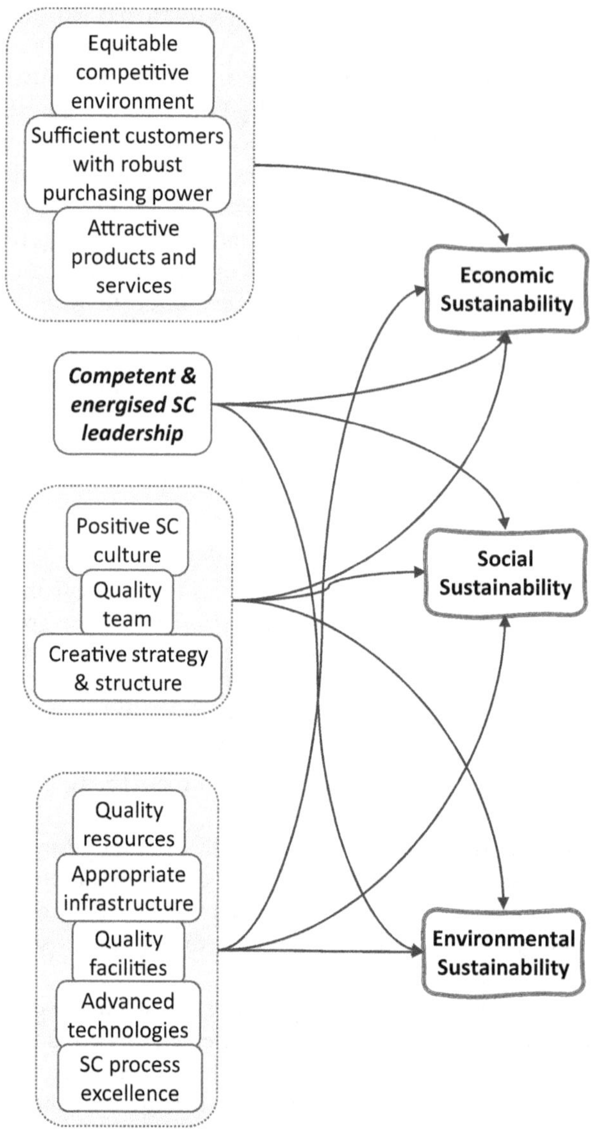

Figure 8.3 Line-of-Sight Between Precondition Clusters and Sustainability Goals

8.4 Chapter 8 case study – building supply chain excellence preconditions in practice

The case study for this chapter is based on research undertaken by Kuei et al. (2005) on the importance of SC quality management on SC performance.

The researchers adopted a fairly wide definition of quality management as a process that includes customer focus, the quality of ICT systems and relationships with suppliers – in other words, a number of the many SC processes and sub-processes covered in this book.

Table 8.2 Results of Hypotheses Testing

Management initiative	*Hypothesis*			
	H1	*H2*	*H3*	*H4*
Six Sigma	✓	✓	✓	×
ERP	✓	✓	✓	✓
Supplier partnerships	✓	✓	✓	✓

The researchers point out that SCs have over time adopted a wide range of improvement methodologies including Lean, Six Sigma, CPFR, ERP, TQM, business process reengineering and various of the ISO standards in order to excel on dimensions important to the longevity of the SC and especially on the dimensions significant to customers. In effect, the researchers were describing SC excellence, albeit with a specific focus on the effectiveness of Six Sigma, ERP and supplier partnerships in delivering SC performance excellence.

The researchers posed two main research questions:

1 What effects do SC quality management initiatives, such as Six Sigma, ERP and supplier partnership, have on SC performance?
2 What trends, if any, influence SC competence?

For each of the three SC quality management initiatives (Six Sigma, ERP and supplier partnerships), four hypotheses were proposed:

1 H1: management $initiative_i$ firms can be discriminated from non-management $initiative_i$ firms by their *SC practices*
2 H2: management $initiative_i$ firms can be discriminated from non-management $initiative_i$ firms by their *application of SC principles*
3 H3: management $initiative_i$ firms can be discriminated from non-management $initiative_i$ firms by their *desired SC capabilities*
4 H4: management $initiative_i$ firms can be discriminated from non-management $initiative_i$ firms by their *organisational performance indicators*

To gather the data necessary to answer the research questions, the researchers sent questionnaires to 250 Hong Kong based organisations participating in recognised SCs. Using the 122 responses received, the researchers conducted stepwise discriminate analysis to identify the indicators that separated the organisations studied for each of Six Sigma (21), non-Six Sigma (101), ERP (96), non-ERP (26), supplier partnerships (19) and non-supplier partnerships (103).

The results of the hypotheses testing are summarised in Table 8.2.

It can be seen that all of the hypotheses are supported except for $H_{1,4}$ indicating that the presence or absence of the management initiatives considered were discriminated in terms of a number of SC competence items. Such competence items included (Kuei et al., 2005):

1 Better understanding of customers' future needs
2 Customer needs responsiveness

3 Reliable customer order fulfilment
4 Response time reductions across the SC
5 Improving the overall SC's JIT capability
6 Active supplier relations management
7 Extending the SC beyond n−1 suppliers and n+1 customers
8 Robust product quality assurance processes
9 Actively scanning market demand signals and responding quickly
10 Product assembly deferral as far down the SC as possible
11 Growing the SC's social performance
12 Reducing the SC's overall environmental footprint
13 Enhancing the SC's integrity
14 Active sharing of relevant SC information
15 Active use of quality data and quality data analysis tools

It is interesting to note that the list of competence items described by the researchers as significant in the SC factors they studied include many of the preconditions and outcomes as described in this chapter. This case study is a good example of the value in carrying out decent research towards developing insights and understandings of the topic under review and exploring how that might fit into the topic's wider field.

8.5 Review questions

i If you were asked to summarise the preconditions necessary to achieve a state of SC excellence, what would be the main elements or element clusters in your summary?
ii Are the elements in your summary of necessary preconditions interrelated in any way? If yes, in what way?
iii Of the *socio*-preconditions described in this chapter, are there any that are more important than the others? If yes, why is this the case in your view?
iv Of the *technical*-preconditions described in this chapter, are there any that are more important than the others? If yes, why is this the case in your view?
v Describe what is meant by an *accountabilities hierarchy*. How do employees go about exercising judgement within predefined boundaries under such a hierarchy?
vi If your boss were to insist that you use an *unproven solution* as the mainstay of a major project you have been asked to head up, what would be your best response?
vii What do you understand a 'black swan' project to be? How can such 'black swan' projects be prevented?
viii Why is it that well-performing SCs are more likely to be successful with SC *sustainability* than poorly performing SCs?
ix What is it about complex systems, such as the interrelationships between SC excellence preconditions and SC sustainability, that SC professionals need to pay particular attention to?

8.6 Assignment topics

(Each to be 3,000 words minimum or about three A4 pages single spaced. Tables and figures are not to be included in the word count.)

i From your own research, describe a SC that was confronted with a definite shock. Describe that nature of the shock (historical frequency, scope, ability to be detected, rate of effect spread, severity of impact) and indicate the level of preparedness of the SC before the shock happened. Also describe the felt impact of the shock environmentally, socially and economically. How did the SC respond to the shock and what were the results of that response?

ii The CPQQRT *task assignment model* can be used as both *a task standard* and a *measure of performance* for an individual who has the responsibility of carrying out the task. Using a worked example, explain and illustrate how this model can be applied in practice. Is it a reliable method for appraising an individual's performance in your judgement?

iii From your own research, describe an organisation that is using a genuine *omni-channel marketing strategy*. How does your chosen organisation go about it – What are the key features? What actual results does your chosen organisation obtain from such a strategy? Is it sustainable do you think? Justify your answer

iv From your own research, identify the characteristics of a 'blue-ocean' competitive environment. Also, identify and describe at least three organisations, or whole SCs, that are operating in such an environment. What is the average return-on-investment (or return-on-capital) that organisations in 'blue-oceans' achieve compared to those operating in 'red-oceans'? Are the preconditions for SC success the same or different in each of the competitive environments? If the same, then why is that the case do you think? If different, in what way are they different?

8.7 Chapter 8 references

Bloch, M., Blumberg, S., Laartz, J. (2012). Delivering Large-Scale IT Projects on-Time, on Budget, and on Value. *McKinsey Digital*. Available at: www.mckinsey.com/business-functions/mckinsey-digital/our-insights/delivering-large-scale-it-projects-on-time-on-budget-and-on-value.

Goldratt, E. (1997). *Critical Chain*. Great Barrington, MA: The North River Press.

Koh, S., Gunasekaran, A., Goodman, T. (2011). Drivers, Barriers and Critical Success Factors for ERPII Implementation in Supply Chains: A Critical Analysis. *Journal of Strategic Information Systems*, 20, pp. 385–402.

Kuei, C., Madu, C., Chow, W., Lu, M. (2005). Supply Chain Quality and Excellence in the New Economy: An Empirical Study of Hong Kong Based Firms. *Multinational Business Review*, 13(1), pp. 33–53.

Lawrence, E. (2019). Key Question in Lawsuit Against FCA: Did Company Knowingly Sell Defective Shifters? *Detroit Free Press*. Available at: www.freep.com/story/money/cars/chrysler/2019/12/10/fiat-chrysler-defective-shifters-class-action/2630714001/.

Pagell, M., Wu, Z. (2009). Building a More Complete Theory of Sustainable Supply Chain Management Using Case Studies of 10 Exemplars. *Journal of Supply Chain Management*, 45(2), pp. 37–56.

Robertson, P. (2021). *Supply Chain Analytics*. Abingdon: Routledge.

Index

For Product Safety Concerns and Information please contact our EU representative GPSR@taylorandfrancis.com
Taylor & Francis Verlag GmbH, Kaufingerstraße 24, 80331 München, Germany

www.ingramcontent.com/pod-product-compliance
Lightning Source LLC
LaVergne TN
LVHW080810011026
849160LV00015B/1587
9780367540081